Counterexamples
in
Calculus

Originally published by Maths Press, New Zealand in 2004. ISBN: 0-476-01215-5.

© 2010 by the Mathematical Association of America, Inc.

Library of Congress Catalog Card Number 2009940147
ISBN 978-0-88385-765-6
Printed in the United States of America
Current Printing (last digit):
10 9 8 7 6 5 4 3 2 1

Counterexamples
in
Calculus

Sergiy Klymchuk

Auckland University of Technology

Published and Distributed by
The Mathematical Association of America

CLASSROOM RESOURCE MATERIALS

Classroom Resource Materials is intended to provide supplementary classroom material for students—laboratory exercises, projects, historical information, textbooks with unusual approaches for presenting mathematical ideas, career information, etc.

MAA Service Center

P.O. Box 91112

Washington, DC 20090-1112

1-800-331-1MAA   FAX: 1-301-206-9789

Foreword

This book offers to students and teachers who know that there is more to learning calculus than solving problems mechanically, a welcome and refreshing antidote to rote learning. It is consistent with views I have put forward with A. Watson in *Mathematics as a Constructive Activity: The Role of Learner-Generated Examples* (Mahwah: Erlbaum, 2005). Mathematics is a constructive activity, and a central aspect of learning mathematics is enriching the space of examples that come to mind when one encounters a technical term. The care and precision needed for using and doing mathematics require access to a wide range of "familiar" and "pathological" examples.

Students who use mathematics in other disciplines often have a cavalier attitude toward conditions and constraints. Desperate to complete a task, they pay scant regard to conditions that are necessary for applying a theorem or technique. Teachers can instill in these students a more mathematical approach to problem solving by training them to search for counterexamples.

Counterexamples in Calculus provides all students a foundation on which to build an upward spiral of understanding of the calculus, and an appreciation for the mathematics behind it.

Professor John Mason
Centre for Mathematics Education
The Open University
United Kingdom

Contents

Introduction

Counterexamples in Calculus is a resource for single-variable calculus courses. The book challenges students to provide counterexamples to carefully constructed *incorrect* mathematical statements. Some of the incorrect statements are converses of well-known theorems. Others come from altering or omitting conditions in theorems, or from applying incorrect definitions. I have grouped the incorrect statements into five sections: Functions, Limits, Continuity, Differential Calculus and Integral Calculus. And I have arranged the statements in each section in order of increasing difficulty, emphasizing early in each section some standard misconceptions. The more challenging statements often *seem* correct to students, who may be hard pressed to understand their subtleties and to find appropriate counterexamples. While the book shows counterexamples to each incorrect statement, I encourage students to construct their own and to compare what they have found to the counterexamples suggested by their peers as well as those shown in the book.

A book with a similar mission is *Counterexamples in Analysis* (Gelbaum & Olmsted, 1964), a well-known resource for advanced calculus and analysis courses. In contrast, *Counterexamples in Calculus* focuses mainly on the mathematics covered in introductory calculus courses. The two books feature different statements and examples and have little overlap. The counterexamples shown here generally appear with their graphs, because the book aims to serve as a learning resource for students as well as a teaching and professional development resource for instructors.

Dealing with counterexamples for the first time can be challenging for students. When they hear they can disprove a wrong statement by provid-

ing one counterexample, many students think they can "prove" a correct
statement by showing an example. Even if they know they cannot prove a
theorem by providing only examples, it is hard for some students to accept
the fact that a single counterexample disproves a statement. Some students
believe that a particular counterexample is just an exception to the rule at
hand, and that no other 'pathological' cases exist. Selden & Selden (1998)
have articulated these ideas:

> Students quite often fail to see a single counterexample as disproving
> a conjecture. This can happen when a counterexample is perceived as
> 'the only one that exists', rather than being seen as generic.

With experience, though, students understand the role of counterexamples and become interested in creating them. Using counterexamples to disprove wrong statements can generate many questions for discussion. What
changes will make the statement at hand correct? How can you change a
counterexample and have it remain one? Can you think of other statements
that your counterexample refutes? Can you find another type of function al-
together that will be a counterexample or construct a general class of counterexamples to the statement at hand?

In developing counterexamples, students are forced to pay attention to
every detail in a statement—the word order, the symbols used, the shape of
brackets defining intervals, whether the statement applies to a point or to an
interval, and so on. Consider the following theorem from first-year calculus.

> If a function $f(x)$ is differentiable on (a, b) and its derivative is posi-
> tive for all x in (a, b), then the function is increasing on (a, b).

The following two statements look quite similar to this theorem, but
both are incorrect:

> If a function $f(x)$ is differentiable on (a, b) and its derivative is posi-
> tive at a point $x = c$ in (a, b), then there is a neighborhood of the point
> $x = c$ where the function is increasing;

> If a function $f(x)$ is differentiable on its domain and its derivative
> is positive for all x from its domain, then the function is increasing
> everywhere on its domain.

Students who find counterexamples to the last two statements must grap-
ple with their subtle differences.

A case study from my experience using counterexamples shows how
they can foster discussion and understanding. Consider Statement 3.14 from
the Continuity chapter.

Statement   If a function $y = f(x)$ is defined on $[a, b]$ and continuous on (a, b), then for any N between $f(a)$ and $f(b)$ there is some point $c \in (a, b)$ such that $f(c) = N$.

The only difference a student sees between this statement and the Intermediate Value Theorem is in the shape of the brackets of the interval where the function is continuous: the function is continuous on an open interval (a, b), instead of a closed interval $[a, b]$. When students are asked to disprove the statement they usually come up with something like this:

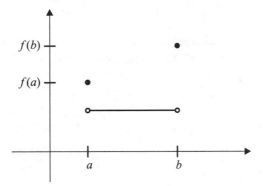

To generate discussion and create other counterexamples one can suggest that:

In the above graph the statement's conclusion is not true for any value of N between $f(a)$ and $f(b)$. Modify the graph in such a way that the statement's conclusion is true for:

a) *exactly one value of N between $f(a)$ and $f(b)$;*

b) *infinitely many but not all values of N between $f(a)$ and $f(b)$.*

One can then expect students to sketch graphs like these:

a)

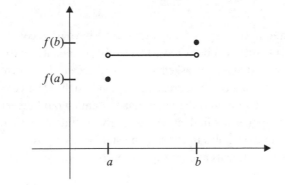

b)

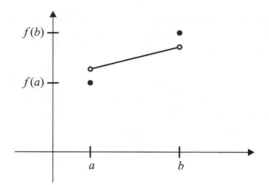

If students are challenged to *give as a counterexample a graph that doesn't have "white circles"*, they may come up with something like this:

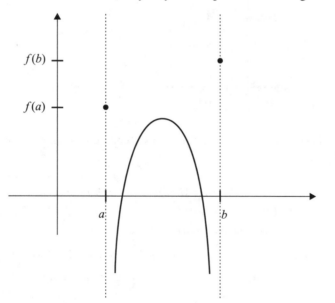

I mention here a few of the other ways I have used counterexamples in teaching. On different occasions, I have given students mixtures of correct and incorrect statements, asked students to create their own wrong statements and associated counterexamples, made deliberate errors in my lecture and moved on with the hope that students would detect them, asked students to spot errors in their textbook, given students extra credit for providing counterexamples to challenging statements I posed in class, and included on assignments and tests questions that require students to construct counterexamples.

Teaching experience has shown that using counterexamples fosters discovery and makes learning more active. In an international study involving more than 600 students from 10 universities in different countries (Gruenwald & Klymchuk, 2003) 92% of the participating students found the use of counterexamples to be very effective. The students reported that it helped them to understand concepts better, prevent mistakes, develop logical and critical thinking, and made learning mathematics more challenging, interesting and creative. Another study (Klymchuk, 2005) showed that the use of counterexamples in teaching improved students' performance on test questions that required conceptual understanding. It is my belief that working with counterexamples reduces misconceptions that can arise in calculus courses which often avoid special cases and expose students only to 'nice' functions and 'good' simple examples, misconceptions explained by Tall's generic extension principle: "If an individual works in a restricted context in which all the examples considered have a certain property, then, in the absence of counterexamples, the mind assumes the known properties to be implicit in other contexts." (Tall, 1991).

Some calculus students may profit from a general introduction to counterexamples, so I have provided some at the beginning of the book. It may be helpful to photocopy these pages for these students before turning them loose on the incorrect statements.

Finally, I would like to express my sincere gratitude and appreciation to the editors and reviewers of the book for their constructive comments and suggestions:

Susan Staples (Texas Christian University) and Jerry Bryce (Hampden-Sydney College) who did the final edit.

Members of the CRM Board who reviewed this book:
Wayne Roberts (Macalester College), Loren Pitt (University of Virginia), Diane Herrmann (University of Chicago), Bill Bauldry (Appalachian State University), Holly Zullo (Carroll College), Phil Straffin (Beloit College).

However, for any inaccuracies, I as the author accept full responsibility.

Counterexamples in Calculus

This book is about counterexamples. Deciding on an assertion's validity is important in the information age. A counterexample can quickly and easily show that a given statement is false. One counterexample is all you need to disprove a statement! Counterexamples thus offer powerful and effective tools for mathematicians, scientists, and researchers. They can indicate that a hypothesis is wrong or a research proposal, misguided. Before attempting a proof for an assertion, looking for counterexamples may save an investigator lots of time and effort.

The search for counterexamples has been important in the history of mathematics. I mention three famous instances. For a long time mathematicians tried to find a formula that would generate only prime numbers. Numbers of the form $2^{2^n} + 1$, where n is a natural number were once believed to all be prime, until Euler found a counterexample. He showed that for $n = 5$ that number is composite:

$$2^{2^5} + 1 = 641 \times 6700417.$$

Primes of this form are called Fermat primes and play an important role in determining which regular n-gons are constructible. (See, for example, Aaboe (1975) p. 84 ff.).

Another conjecture about prime numbers is still awaiting proof or disproof. The Goldbach-Euler conjecture, posed by Christian Goldbach in a 1742 letter to Euler, looks deceptively simple: *every even integer greater than 2 is the sum of two prime numbers.* For example, $12 = 5 + 7$, $20 = 3 + 17$, and so on. Powerful computers have been used to search for possible counterexamples and have found none among $4, 6, \ldots, 4 \times 10^{14}$ (see Rich-

stein, 2000). In 2000 the book publishing company Faber & Faber offered
a \$1 million prize to anyone who could prove or disprove this conjecture
within two years. The prize has so far gone unclaimed.

In the nineteenth century the great German mathematician Karl Weierstrass constructed his famous counterexamples to this statement: *If a function is continuous on (a, b), then it is differentiable at some points on (a, b).*
Functions of this type are introduced in a later chapter. Weierstrass first
showed his counterexamples in lectures in 1861 and later published them in
a paper in 1872, but Bernhard Bolzano had found one in about 1830.) (See
Boyer, 1991, p. 604 ff.)

In this book I ask you to disprove a host of incorrect statements by find-
ing counterexamples. How should you search for appropriate counterexamples? In many cases a simple sketch of a graph is enough. For example,
you might sketch a smooth graph with a vertical tangent to disprove the
statement "*If a function is continuous on the interval (a, b) and its graph
is a smooth curve (no sharp corners) on that interval, then the function is
differentiable at any point on (a, b).*"

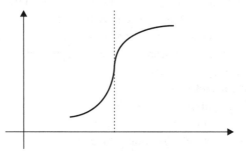

If your arsenal is big enough, you can provide a formula *and* a graph.

$$y = \sqrt[3]{x}$$

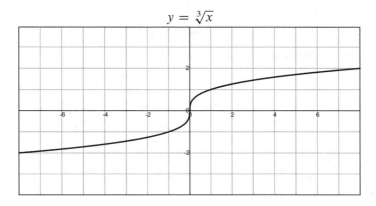

You will learn that in creating counterexamples it is useful to have at your disposal a large assortment of graphs and functions with interesting properties. Understanding the anomalies and distinguishing features of these functions will provide you with a natural starting point for developing your own counterexamples. In your searches you should consider graphs of basic trigonometric functions and their inverses, graphs of piecewise functions like step functions, and graphs with sharp corners (like that of the absolute value function $|x|$) or cusps ($\sqrt[3]{x^2}$ is a handy example). The quest for counterexamples will introduce you quite naturally to more exotic functions, like oscillations with and without damping factors (such as $\sin\frac{1}{x}$, $x\sin\frac{1}{x}$, and $x^2|\cos\frac{\pi}{x}|$), and to even more exotic functions like Dirichlet and Weierstrass functions, which are notable in the history of mathematics.

So get ready to create counterexamples! Have fun disproving the statements!

Part I

Statements

1

Functions

1.1 The tangent to a curve at a point is the line that touches the curve at that point, but does not cross it there.

1.2 The tangent line to a curve at a point cannot touch the curve at infinitely many other points.

1.3 A quadratic function of x is one in which the highest power of x is two.

1.4 If both functions $f(x)$ and $g(x)$ are continuous and monotone on $\mathbb{R}$, then their sum $f(x) + g(x)$ is also monotone on $\mathbb{R}$.

1.5 If both functions $f(x)$ and $g(x)$ are not monotone on $\mathbb{R}$, then their sum $f(x) + g(x)$ is not monotone on $\mathbb{R}$.

1.6 If a function $f(x)$ is continuous and decreasing for all positive x and $f(1)$ is positive, then the function has exactly one root.

1.7 If a function $f(x)$ has an inverse function $f^{-1}(y)$ on (a, b), then the function $f(x)$ is either increasing or decreasing on (a, b).

1.8 A function $f(x)$ is bounded on $\mathbb{R}$ if for each $x \in \mathbb{R}$ there is $M > 0$ such that $|f(x)| \leq M$.

1.9 If $g(a) = 0$, then the function

$$F(x) = \frac{f(x)}{g(x)}$$

has a vertical asymptote at the point $x = a$.

1.10 If $g(a) = 0$, then the *rational* function

$$R(x) = \frac{f(x)}{g(x)}$$

(both $f(x)$ and $g(x)$ are polynomials) has a vertical asymptote at the point $x = a$.

1.11 If a function $f(x)$ is unbounded and nonnegative for all real x, then it cannot have roots x_n such that $x_n \to \infty$ as $n \to \infty$.

1.12 A function $f(x)$ defined on $[a, b]$ such that its graph does not contain any pieces of a horizontal straight line cannot take its extreme value infinitely many times on $[a, b]$.

1.13 If a function $f(x)$ is continuous and increasing at the point $x = a$, then there is a neighborhood $(x - \delta, x + \delta)$, $\delta > 0$ where the function is also increasing.

1.14 If a function is not monotone, then it does not have an inverse function.

1.15 If a function is not monotone on (a, b), then its square cannot be monotone on (a, b).

2

Limits

2.1 If $f(x) < g(x)$ for all $x > 0$ and both

$$\lim_{x \to \infty} f(x) \quad \text{and} \quad \lim_{x \to \infty} g(x)$$

exist, then

$$\lim_{x \to \infty} f(x) < \lim_{x \to \infty} g(x).$$

2.2 The following definitions of a non-vertical asymptote are equivalent:

a) The straight line $y = mx + c$ is called a non-vertical asymptote to a curve $f(x)$, as x tends to infinity, if

$$\lim_{x \to \infty} \left(f(x) - (mx + c) \right) = 0.$$

b) A straight line is called a non-vertical asymptote to a curve, as x tends to infinity, if the curve gets closer and closer (as close as we like) to the straight line as x tends to infinity without touching or crossing it.

2.3 The tangent line to a curve at a certain point that touches the curve at infinitely many other points cannot be a non-vertical asymptote to this curve.

2.4 The following definitions of a vertical asymptote are equivalent:

15

a) The straight line $x = a$ is called a vertical asymptote for a function $f(x)$ if

$$\lim_{x \to a^+} f(x) = \pm\infty \quad \text{or} \quad \lim_{x \to a^-} f(x) = \pm\infty.$$

b) The straight line $x = a$ is called a vertical asymptote for the function $f(x)$ if there are infinitely many values of $f(x)$ that can be made arbitrarily large or arbitrarily small as x gets closer to a from either side of a.

2.5 If $\lim_{x \to a} f(x)$ exists and $\lim_{x \to a} g(x)$ does not exist because of oscillation of $g(x)$ near $x = a$, then $\lim_{x \to a} \big(f(x) g(x)\big)$ does not exist.

2.6 If a function $f(x)$ is not bounded in any neighborhood of the point $x = a$, then either

$$\lim_{x \to a^+} |f(x)| = \infty \quad \text{or} \quad \lim_{x \to a^-} |f(x)| = \infty.$$

2.7 If a function $f(x)$ is continuous for all real x and $\lim_{n \to \infty} f(n) = A$ for natural numbers n, then $\lim_{x \to \infty} f(x) = A$.

3

Continuity

3.1 If the absolute value of the function $f(x)$ is continuous on (a, b), then the function is also continuous on (a, b).

3.2 If both functions $f(x)$ and $g(x)$ are discontinuous at $x = a$, then $f(x) + g(x)$ is also discontinuous at $x = a$.

3.3 If both functions $f(x)$ and $g(x)$ are discontinuous at $x = a$, then $f(x) g(x)$ is also discontinuous at $x = a$.

3.4 A function always has a local maximum between any two local minima.

3.5 For a continuous function there is always a strict local maximum between any two local minima.

3.6 If a function is defined in a certain neighborhood of point $x = a$ including the point itself and is increasing for all $x < a$ and decreasing for all $x > a$, then there is a local maximum at $x = a$.

3.7 If a function is defined on $[a, b]$ and continuous on (a, b), then it takes its extreme values on $[a, b]$.

3.8 Every continuous and bounded function on $(-\infty, \infty)$ takes on its extreme values.

3.9 If a function $f(x)$ is continuous on $[a, b]$, the tangent line exists at all points on its graph and $f(a) = f(b)$, then there is a point c in (a, b) such that the tangent line at the point $(c, f(c))$ is horizontal.

3.10 If on the closed interval $[a, b]$ a function is:

   a) bounded;

   b) takes its maximum and minimum values;

   c) takes all its values between the maximum and minimum values;

then this function is continuous on [a,b].

3.11 If on the closed interval $[a, b]$ a function is:

   a) bounded;

   b) takes its maximum and minimum values;

   c) takes all its values between the maximum and minimum values;

then this function is continuous at one or more points or subintervals on $[a, b]$.

3.12 If a function is continuous on $[a, b]$, then it cannot take its absolute maximum or minimum value infinitely many times.

3.13 If a function $f(x)$ is defined on $[a, b]$ and $f(a) \times f(b) < 0$, then there is some point $c \in (a, b)$ such that $f(c) = 0$.

3.14 If a function $f(x)$ is defined on $[a, b]$ and continuous on (a, b), then for any $N \in (f(a), f(b))$ there is some point $c \in (a, b)$ such that $f(c) = N$.

3.15 If a function is discontinuous at every point in its domain, then the square and the absolute value of this function cannot be continuous.

3.16 A function cannot be continuous at only one point in its domain and discontinuous everywhere else.

3.17 A sequence of continuous functions on $[a, b]$ always converges to a continuous function on $[a, b]$.

4

Differential Calculus

4.1 If both functions $f(x)$ and $g(x)$ are differentiable and $f(x) > g(x)$ on the interval (a, b), then $f'(x) > g'(x)$ on (a, b).

4.2 If a nonlinear function is differentiable and monotone on $(0, \infty)$, then its derivative is also monotone on $(0, \infty)$.

4.3 If a function is continuous at a point, then it is differentiable at that point.

4.4 If a function is continuous on $\mathbb{R}$ and the tangent line exists at any point on its graph, then the function is differentiable at any point on $\mathbb{R}$.

4.5 If a function is continuous on the interval (a, b) and its graph is a *smooth* curve (no sharp corners) on that interval, then the function is differentiable at any point on (a, b).

4.6 If the derivative of a function is zero at a point, then the function is neither increasing nor decreasing at this point.

4.7 If a function is differentiable and decreasing on (a, b), then its derivative is negative on (a, b).

4.8 If a function is continuous and decreasing on (a, b), then its derivative is nonpositive on (a, b).

4.9 If a function has a positive derivative at every point in its domain, then the function is increasing everywhere in its domain.

4.10 If a function $f(x)$ is defined on $[a, b]$ and has a local maximum at the point $c \in (a, b)$, then in a sufficiently small neighborhood of the point $x = c$, the function is increasing for all $x < c$ and decreasing for all $x > c$.

4.11 If a function $f(x)$ is differentiable for all real x and $f(0) = f'(0) = 0$, then $f(x) = 0$ for all real x.

4.12 If a function $f(x)$ is differentiable on the interval (a, b) and takes both positive and negative values on it, then its absolute value $|f(x)|$ is not differentiable at the point(s) where $f(x) = 0$, e.g.,

$$|f(x)| = |x| \quad \text{or} \quad |f(x)| = |\sin x|.$$

4.13 If both functions $f(x)$ and $g(x)$ are differentiable on the interval (a, b) and intersect somewhere on (a, b), then the function

$$\max \{f(x), g(x)\}$$

is not differentiable at the point(s) where $f(x) = g(x)$.

4.14 If a function is twice-differentiable at a local maximum (minimum) point, then its second derivative is negative (positive) at that point.

4.15 If both functions $f(x)$ and $g(x)$ are not differentiable at $x = a$, then $f(x) + g(x)$ is also not differentiable at $x = a$.

4.16 If a function $f(x)$ is differentiable and a function $g(x)$ is not differentiable at $x = a$, then $f(x) g(x)$ is not differentiable at $x = a$.

4.17 If both functions $f(x)$ and $g(x)$ are not differentiable at $x = a$, then $f(x) g(x)$ is also not differentiable at $x = a$.

4.18 If a function $g(x)$ is differentiable at $x = a$ and a function $f(x)$ is not differentiable at $g(a)$, then the function $F(x) = f(g(x))$ is not differentiable at $x = a$.

4.19 If a function $g(x)$ is not differentiable at $x = a$ and a function $f(x)$ is differentiable at $g(a)$, then the function $F(x) = f\big(g(x)\big)$ is not differentiable at $x = a$.

4.20 If a function $g(x)$ is not differentiable at $x = a$ and a function $f(x)$ is not differentiable at $g(a)$, then the function $F(x) = f\big(g(x)\big)$ is not differentiable at $x = a$.

4.21 If a function $f(x)$ is defined on $[a, b]$, differentiable on (a, b) and $f(a) = f(b)$, then there exists a point $c \in (a, b)$ such that $f'(c) = 0$.

4.22 If a function is twice-differentiable in a certain neighborhood of the point $x = a$ and its second derivative is zero at that point, then the point $\big(a, f(a)\big)$ is a point of inflection for the graph of the function.

4.23 If a function $f(x)$ is differentiable at the point $x = a$ and the point $\big(a, f(a)\big)$ is a point of inflection on the function's graph, then the second derivative is zero at that point.

4.24 If both functions $f(x)$ and $g(x)$ are differentiable on $\mathbb{R}$, then to evaluate the limit $\lim_{x \to \infty} \frac{f(x)}{g(x)}$ in the indeterminate form of type $\left[\frac{\infty}{\infty}\right]$ we can use the following rule:

$$\lim_{x \to \infty} \frac{f(x)}{g(x)} = \lim_{x \to \infty} \frac{f'(x)}{g'(x)}.$$

4.25 If a function $f(x)$ is differentiable on (a, b) and $\lim_{x \to a+} f'(x) = \infty$, then $\lim_{x \to a+} f(x) = \infty$.

4.26 If a function f(x) is differentiable on $(0, \infty)$ and $\lim_{x \to \infty} f(x)$ exists, then $\lim_{x \to \infty} f'(x)$ also exists.

4.27 If a function $f(x)$ is differentiable and bounded on $(0, \infty)$ and $\lim_{x \to \infty} f'(x)$ exists, then $\lim_{x \to \infty} f(x)$ also exists.

4.28 If a function $f(x)$ is differentiable at the point $x = a$, then its derivative is continuous at $x = a$.

4.29 If the derivative of a function $f(x)$ is positive at the point $x = a$, then there exists a neighborhood about $x = a$ where the function is increasing.

4.30 If a function $f(x)$ is continuous on (a, b) and has a local maximum at the point $c \in (a, b)$, then in a sufficiently small neighborhood of the point $x = c$ the function is increasing for all $x < c$ and decreasing for all $x > c$.

4.31 If a function $f(x)$ is differentiable at the point $x = a$, then there is a certain neighborhood of the point $x = a$ where the derivative of the function $f(x)$ is bounded.

4.32 If a function f(x) in every neighborhood of the point x = a has points where $f'(x)$ does not exist, then $f'(a)$ does not exist.

4.33 A function cannot be differentiable only at one point in its domain and nondifferentiable everywhere else in its domain.

4.34 If a function is continuous on (a,b), then it is differentiable at some points on (a,b).

5

Integral Calculus

5.1 If the function $F(x)$ is an antiderivative of a function $f(x)$, then

$$\int_a^b f(x)\,dx = F(b) - F(a).$$

5.2 If a function $f(x)$ is continuous on $[a, b]$, then the area enclosed by the graph of $y = f(x)$, $y = 0$, $x = a$ and $x = b$ numerically equals

$$\int_a^b f(x)\,dx.$$

5.3 If

$$\int_a^b f(x)\,dx \geq 0,$$

then $f(x) \geq 0$ for all $x \in [a, b]$.

5.4 If $f(x)$ is a continuous function and k is any constant, then:

$$\int k f(x)\,dx = k \int f(x)\,dx.$$

5.5 A plane figure of an infinite area rotated about an axis always produces a solid of revolution of infinite volume.

23

5.6 If a function $f(x)$ is defined for every $x \in [a, b]$ and

$$\int_a^b |f(x)|\, dx$$

exists, then

$$\int_a^b f(x)\, dx$$

exists.

5.7 If neither of the integrals

$$\int_a^b f(x)\, dx \quad \text{and} \quad \int_a^b g(x)\, dx$$

exist, then the integral

$$\int_a^b \left(f(x) + g(x)\right) dx$$

does not exist.

5.8 If $\lim_{x \to \infty} f(x) = 0$, then

$$\int_a^\infty f(x)\, dx$$

converges.

5.9 If the integral

$$\int_a^\infty f(x)\, dx$$

diverges, then the function $f(x)$ is not bounded.

5.10 If a function $f(x)$ is continuous and nonnegative for all real x and $\sum_{n=1}^\infty f(n)$ is finite, then

$$\int_1^\infty f(x)\, dx$$

converges.

5.11 If both integrals

$$\int_a^\infty f(x)\, dx \quad \text{and} \quad \int_a^\infty g(x)\, dx$$

diverge, then the integral

$$\int_a^\infty (f(x) + g(x))\, dx$$

also diverges.

5.12 If a function $f(x)$ is continuous and

$$\int_a^\infty f(x)\, dx$$

converges, then $\lim_{x \to \infty} f(x) = 0$.

5.13 If a function $f(x)$ is continuous and nonnegative and

$$\int_a^\infty f(x)\, dx$$

converges, then $\lim_{x \to \infty} f(x) = 0$.

5.14 If a function $f(x)$ is positive and unbounded for all real x, then the integral

$$\int_a^\infty f(x)\, dx$$

diverges.

5.15 If a function $f(x)$ is continuous and unbounded for all real x, then the integral

$$\int_a^\infty f(x)\, dx$$

diverges.

5.16 If a function $f(x)$ is continuous on $[1, \infty)$ and

$$\int_1^\infty f(x)\, dx$$

converges, then

$$\int_1^\infty |f(x)|\, dx$$

also converges.

5.17 If the integral

$$\int_a^\infty f(x)\,dx$$

converges and a function $g(x)$ is bounded, then the integral

$$\int_a^\infty f(x)g(x)\,dx$$

converges.

Part II

Suggested Solutions

1

Functions

1.1 The tangent to a curve at a point is the line that touches the curve at that point, but does not cross it there.

Counterexample

a) The x-axis is the tangent line to the curve $y = x^3$, but it crosses the curve at the origin.

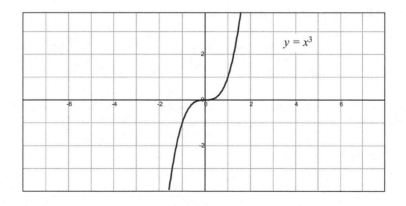

b) The three straight lines just touch and do not cross the curve below at the point, but none of them is the tangent line to the curve at that point.

1.2   The tangent line to a curve at a point cannot touch the curve at infinitely many other points.

Counterexample   The tangent line to the graph of the function $y = \sin x$ touches the curve at $x = \frac{\pi}{2}$ and infinitely many other points.

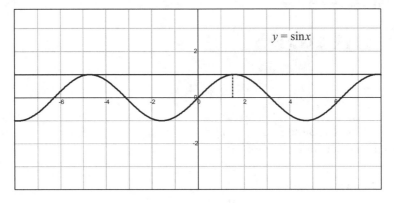

1.3   A quadratic function of x is one in which the highest power of x is two.

Counterexample   In both functions

$$y = x^2 + \sqrt{x} \quad \text{and} \quad y = x^2 + x - \frac{1}{x}$$

the highest power of x is two, but neither is quadratic.

1.4   If both functions $f(x)$ and $g(x)$ are continuous and monotone on $\mathbb{R}$, then their sum $f(x) + g(x)$ is also monotone on $\mathbb{R}$.

Counterexample

$$f(x) = x + \sin x,$$
$$g(x) = -x.$$

Both functions $f(x)$ and $g(x)$ are monotone on $\mathbb{R}$, but their sum $f(x) + g(x) = \sin x$ is not monotone on $\mathbb{R}$.

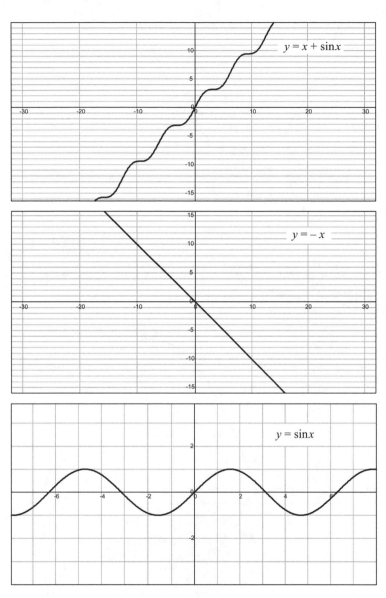

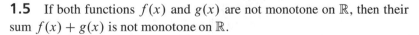

1.5 If both functions $f(x)$ and $g(x)$ are not monotone on $\mathbb{R}$, then their sum $f(x) + g(x)$ is not monotone on $\mathbb{R}$.

Counterexample Both functions $f(x) = x + x^2$ and $g(x) = x - x^2$ are not monotone on $\mathbb{R}$, but their sum $f(x) + g(x) = 2x$ is monotone on $\mathbb{R}$.

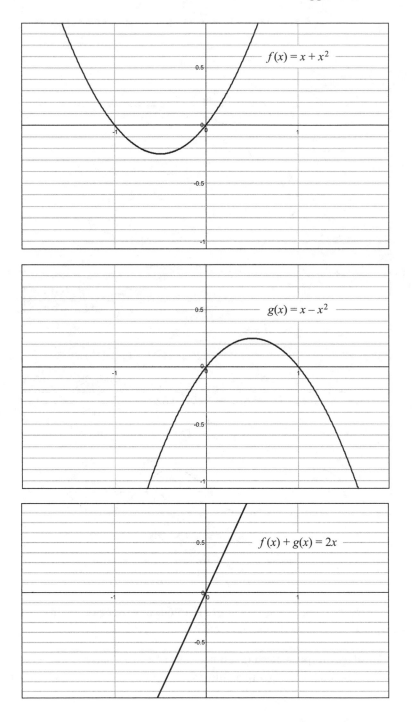

1.6   If a function $f(x)$ is continuous and decreasing for all positive x and $f(1)$ is positive, then the function has exactly one root.

Counterexample   The function $y = \frac{1}{x}$ is continuous and decreasing for all positive x and $y(1) = 1 > 0$, but it has no roots.

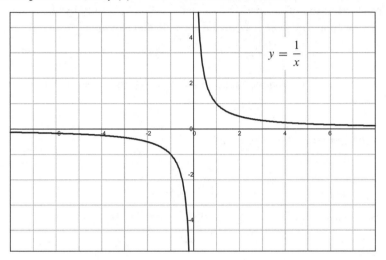

1.7   If a function $f(x)$ has an inverse function $f^{-1}(y)$ on (a, b), then the function $f(x)$ is either increasing or decreasing on (a, b).

Counterexample   The function below is a one-to-one function and has an inverse function on (a, b), but it is neither increasing nor decreasing on (a, b).

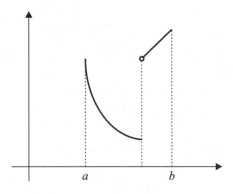

1.8   A function $f(x)$ is bounded on $\mathbb{R}$ if for each $x \in \mathbb{R}$ there is $M > 0$ such that $|f(x)| \leq M$.

Counterexample   For the function $y = x^2$, for any value of x from $\mathbb{R}$, there is a number $M > 0$ ($M = x^2 + \varepsilon$, where $\varepsilon \geq 0$) such that $|f(x)| \leq M$.

Comment   The order of words in this statement is very important. The correct definition of a function bounded on $\mathbb{R}$ differs only by the order of words: A function $f(x)$ is bounded on $\mathbb{R}$ if there is $M > 0$ such that for any $x \in \mathbb{R}$, $|f(x)| \leq M$.

1.9   If $g(a) = 0$, then the function $F(x) = \frac{f(x)}{g(x)}$ has a vertical asymptote at the point $x = a$.

Counterexample   The function

$$y = \frac{\sin x}{x}$$

does not have a vertical asymptote at the point $x = 0$.

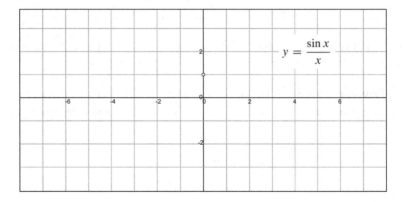

1.10   If $g(a) = 0$, then the *rational* function $R(x) = \frac{f(x)}{g(x)}$ (both $f(x)$ and $g(x)$ are polynomials) has a vertical asymptote at the point $x = a$.

Counterexample   The rational function

$$y = \frac{x^2 - 1}{x - 1}$$

does not have a vertical asymptote at the point $x = 1$.

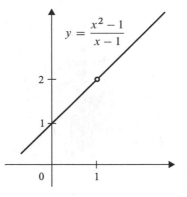

1.11   If a function $f(x)$ is unbounded and nonnegative for all real x, then it cannot have roots x_n such that $x_n \to \infty$ as $n \to \infty$.

Counterexample   The function $y = |x \sin x|$ has infinitely many roots x_n such that $x_n \to \infty$ as $n \to \infty$.

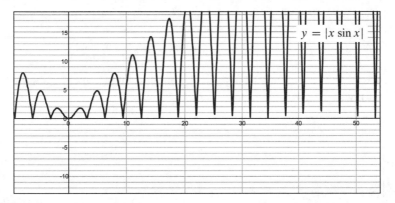

1.12   A function $f(x)$ defined on $[a, b]$ such that its graph does not contain any pieces of a horizontal straight line cannot take its extreme value infinitely many times on $[a, b]$.

Counterexample   The function

$$y = \begin{cases} \sin \frac{1}{x}, & \text{if } x \neq 0 \\ 0, & \text{if } x = 0 \end{cases}$$

takes its absolute maximum value ($= 1$) and its absolute minimum value ($= -1$) infinitely many times on any closed interval containing zero.

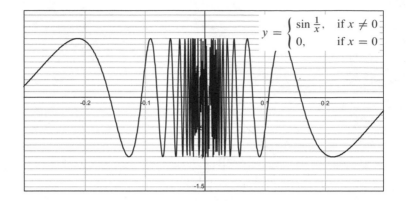

1.13   If a function $f(x)$ is continuous and increasing at the point $x = a$, then there is a neighborhood $(x - \delta, x + \delta)$, $\delta > 0$ where the function is also increasing.

Counterexample   The function

$$f(x) = \begin{cases} x + x^2 \sin \frac{2}{x}, & \text{if } x \neq 0 \\ 0, & \text{if } x = 0 \end{cases}$$

is increasing at the point $x = 0$, but it is not increasing in any neighborhood $(-\delta, \delta)$, where $\delta > 0$.

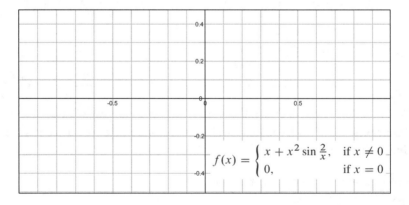

Comment   The definition of a function increasing at a point is: A function $f(x)$ is said to be *increasing at the point* $x = a$ if in a certain neighborhood $(a - \delta, a + \delta)$, $\delta > 0$ the following is true:

if $x < a$ then $f(x) < f(a)$ and if $x > a$ then $f(x) > f(a)$.

1.14   If a function is not monotone, then it does not have an inverse function.

Counterexample   The function

$$y = \begin{cases} x, & \text{if } x \text{ is rational} \\ -x, & \text{if } x \text{ is irrational} \end{cases}$$

is not monotone, but it has the inverse function

$$x = \begin{cases} y, & \text{if } y \text{ is rational} \\ -y, & \text{if } y \text{ is irrational.} \end{cases}$$

It is impossible to draw the graph of such a function, but a rough sketch gives an idea of its behavior:

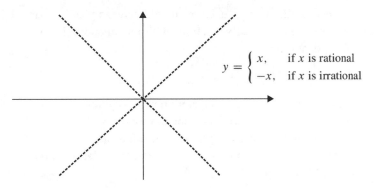

$$y = \begin{cases} x, & \text{if } x \text{ is rational} \\ -x, & \text{if } x \text{ is irrational} \end{cases}$$

1.15   If a function is not monotone on (a, b), then its square cannot be monotone on (a, b).

Counterexample   The function

$$f(x) = \begin{cases} x, & \text{if } x \text{ is rational} \\ -x, & \text{if } x \text{ is irrational} \end{cases}$$

defined on $(0, \infty)$ is not monotone, but its square $f^2(x) = x^2$ is monotone on $(0, \infty)$.

It is impossible to draw the graph of the function $f(x)$, but the sketch below gives an idea of its behavior.

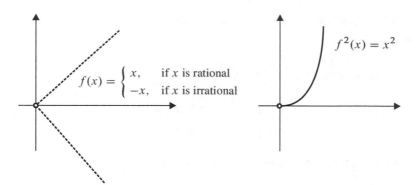

$$f(x) = \begin{cases} x, & \text{if } x \text{ is rational} \\ -x, & \text{if } x \text{ is irrational} \end{cases}$$

$$f^2(x) = x^2$$

Comment   The functions in counterexamples 1.14 and 1.15 may seem artificial and without practical use at first. Nevertheless, the Dirichlet function

$$f(x) = \begin{cases} 1, & \text{if } x \text{ is rational} \\ 0, & \text{if } x \text{ is irrational} \end{cases},$$

which is very similar to the functions in counterexamples 1.14 and 1.15, can be represented analytically as a limit of cosine functions that have many practical applications:

$$f(x) = \lim_{k \to \infty} \lim_{n \to \infty} \left(\cos(k!\pi x) \right)^{2n}.$$

The Dirichlet like functions are not continuous at any point of their domain. They are called nowhere continuous functions or everywhere discontinuous functions. Later in the book you will see more functions like this one used as counterexamples. A good introduction to the Dirichlet function is given in Dunham (2005) on p.197.

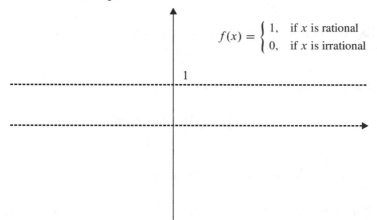

$$f(x) = \begin{cases} 1, & \text{if } x \text{ is rational} \\ 0, & \text{if } x \text{ is irrational} \end{cases}$$

2

Limits

2.1 If $f(x) < g(x)$ for all $x > 0$ and both $\lim\limits_{x \to \infty} f(x)$ and $\lim\limits_{x \to \infty} g(x)$ exist, then $\lim\limits_{x \to \infty} f(x) < \lim\limits_{x \to \infty} g(x)$.

Counterexample For the functions

$$f(x) = -\frac{1}{x} \quad \text{and} \quad g(x) = \frac{1}{x},$$

$f(x) < g(x)$ for all $x > 0$, but

$$\lim_{x \to \infty} f(x) = \lim_{x \to \infty} g(x) = 0.$$

2.2   The following definitions of a non-vertical asymptote are equivalent:

a) The straight line $y = mx + c$ is called a non-vertical asymptote to a curve $f(x)$ as x tends to infinity if $\lim\limits_{x \to \infty} (f(x) - (mx + c)) = 0$.

b) A straight line is called a non-vertical asymptote to a curve as x tends to infinity if the curve gets closer and closer to the straight line (as close as we like) as x tends to infinity, but does not touch or cross it.

Counterexample   As x tends to infinity the function $y = \dfrac{\sin x}{x}$ gets closer to the x-axis from above and below and $\lim\limits_{x \to \infty} \left(\dfrac{\sin x}{x} - 0 \right) = 0$. According to the first definition the x-axis is the non-vertical asymptote of the function $y = \dfrac{\sin x}{x}$, but its graph crosses the x-axis infinitely many times, so the definitions a) and b) are not equivalent.

Comment   The correct definition is a). A function's graph can touch or cross a non-vertical asymptote.

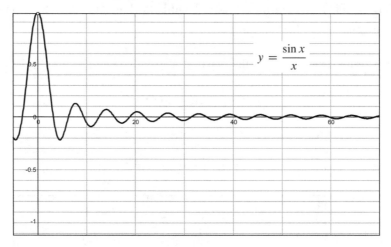

2.3   The tangent line to a curve at a certain point that touches the curve at infinitely many other points cannot be a non-vertical asymptote to this curve.

Counterexample   The tangent line $y = 0$ to the curve $y = \dfrac{\sin^2 x}{x}$ at $x = \pi$ touches the curve at infinitely many other points and is a non-vertical asymptote to this curve.

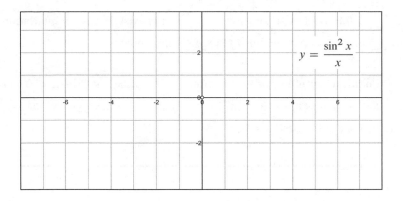

2.4   The following definitions of a vertical asymptote are equivalent:

a) The straight line $x = a$ is called a vertical asymptote for a function $y = f(x)$ if $\lim\limits_{x \to a^+} f(x) = \pm\infty$ or $\lim\limits_{x \to a^-} f(x) = \pm\infty$.

b) The straight line $x = a$ is called a vertical asymptote for the function $f(x)$ if there are infinitely many values of $f(x)$ that can be made arbitrarily large or arbitrarily small as x gets closer to a from either side of a.

Counterexample   There are infinitely many values of the function

$$y = \frac{1}{x} \sin \frac{1}{x}$$

that can be made arbitrarily large or small as x gets closer to 0, but the straight line $x = 0$ is not a vertical asymptote of this function.

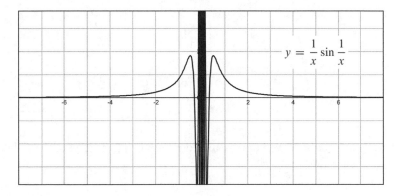

Comment   The correct definition is a).

2.5 If $\lim\limits_{x \to a} f(x)$ exists and $\lim\limits_{x \to a} g(x)$ does not exist because of oscillation of $g(x)$ near $x = a$, then $\lim\limits_{x \to a} \big(f(x) \, g(x) \big)$ does not exist.

Counterexample For the function $f(x) = x$ the limit $\lim\limits_{x \to 0} x = 0$ and for the function $g(x) = \sin \frac{1}{x}$ the limit $\lim\limits_{x \to 0} \sin \frac{1}{x}$ does not exist because of oscillation of $g(x)$ near $x = 0$, but

$$\lim_{x \to 0} \big(f(x) \, g(x) \big) = \lim_{x \to 0} \left(x \sin \frac{1}{x} \right) = 0.$$

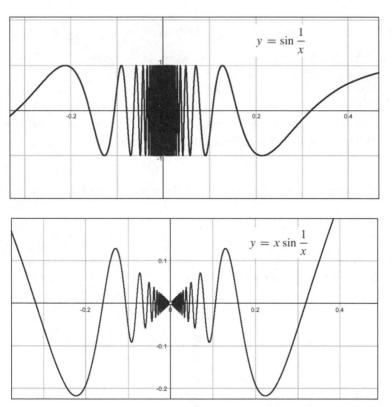

2.6 If a function $f(x)$ is not bounded in any neighborhood of the point $x = a$, then either

$$\lim_{x \to a+} \big| f(x) \big| = \infty \quad \text{or} \quad \lim_{x \to a-} \big| f(x) \big| = \infty.$$

Counterexample   The function

$$f(x) = \frac{1}{x}\cos\frac{1}{x}$$

is not bounded in any neighborhood of the point $x = 0$, but neither

$$\lim_{x \to 0^+} \left| \frac{1}{x}\cos\frac{1}{x} \right| \quad \text{nor} \quad \lim_{x \to 0^-} \left| \frac{1}{x}\cos\frac{1}{x} \right|$$

exist.

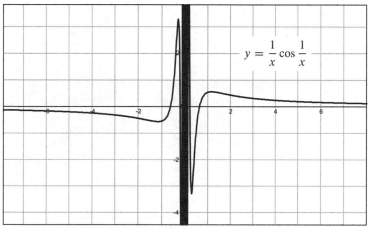

$$y = \frac{1}{x}\cos\frac{1}{x}$$

2.7   If a function $f(x)$ is continuous for all real x and $\lim_{n \to \infty} f(n) = A$ for natural numbers n, then $\lim_{x \to \infty} f(x) = A$.

Counterexample   For the continuous function $y = \cos(2\pi x)$ the limit $\lim_{n \to \infty} \cos(2\pi n)$ equals 1 because $\cos(2\pi n) = 1$ for any natural n, but $\lim_{x \to \infty} \cos(2\pi x)$ does not exist.

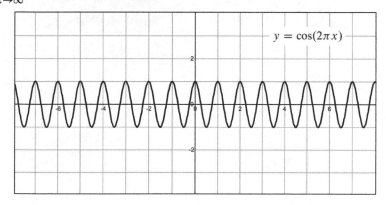

$$y = \cos(2\pi x)$$

Comment   Statement 2.7 is the converse of the true statement:

$$\lim_{x \to \infty} f(x) = A \Rightarrow \lim_{n \to \infty} f(n) = A.$$

3

Continuity

3.1 If the absolute value of the function $f(x)$ is continuous on (a, b), then the function is also continuous on (a, b).

Counterexample   The absolute value of the function

$$y(x) = \begin{cases} -1, & \text{if } x \le 0 \\ 1, & \text{if } x > 0 \end{cases}$$

is $|y(x)| = 1$ for all real x and it is continuous, but the function $y(x)$ is discontinuous.

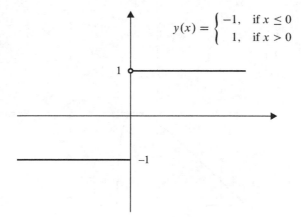

$$y(x) = \begin{cases} -1, & \text{if } x \le 0 \\ 1, & \text{if } x > 0 \end{cases}$$

3.2 If both functions $f(x)$ and $g(x)$ are discontinuous at $x = a$, then $f(x) + g(x)$ is also discontinuous at $x = a$.

Counterexample

$$f(x) = -\frac{1}{x-a}, \quad \text{if } x \neq a$$

$$g(x) = x + \frac{1}{x-a}, \quad \text{if } x \neq a$$

$$f(x) = g(x) = \frac{a}{2}, \quad \text{if } x = a$$

Both functions $f(x)$ and $g(x)$ are discontinuous at $x = a$, but the function

$$f(x) + g(x) = \begin{cases} x, & \text{if } x \neq a \\ a, & \text{if } x = a \end{cases}$$

is continuous at $x = a$. For example, if $a = 2$:

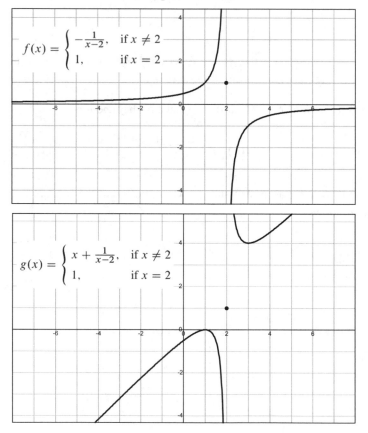

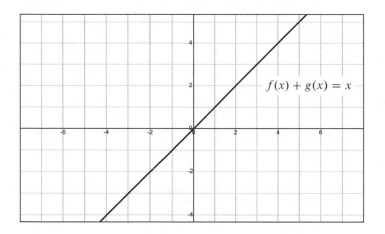

3.3 If both functions $f(x)$ and $g(x)$ are discontinuous at $x = a$, then $f(x) g(x)$ is also discontinuous at $x = a$.

Counterexample Both functions

$$f(x) = \begin{cases} \frac{\sin x}{x}, & \text{if } x \neq 0 \\ 2, & \text{if } x = 0 \end{cases} \quad \text{and} \quad g(x) = \begin{cases} \frac{\sin x}{x}, & \text{if } x \neq 0 \\ \frac{1}{2}, & \text{if } x = 0 \end{cases}$$

are discontinuous at the point $x = 0$, but their product

$$f(x) g(x) = \begin{cases} \frac{\sin^2 x}{x^2}, & \text{if } x \neq 0 \\ 1, & \text{if } x = 0 \end{cases}$$

is continuous at the point $x = 0$.

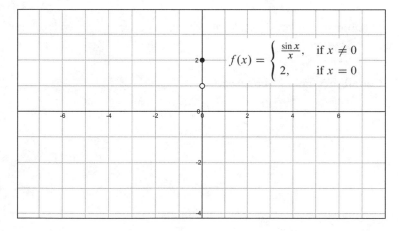

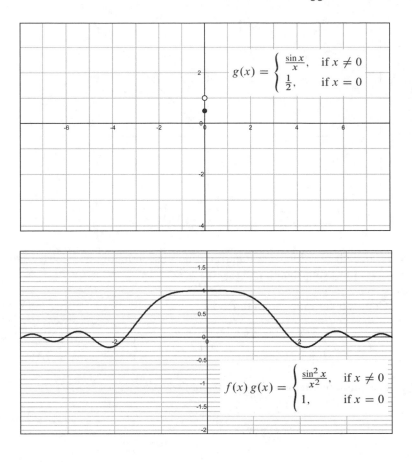

$$g(x) = \begin{cases} \frac{\sin x}{x}, & \text{if } x \neq 0 \\ \frac{1}{2}, & \text{if } x = 0 \end{cases}$$

$$f(x)\,g(x) = \begin{cases} \frac{\sin^2 x}{x^2}, & \text{if } x \neq 0 \\ 1, & \text{if } x = 0 \end{cases}$$

3.4   A function always has a local maximum between any two local minima.

Counterexample   The functions

$$y = \frac{x^4 + 0.1}{x^2}$$

and

$$y = \sec^2 x$$

have no maximum between two local minima:

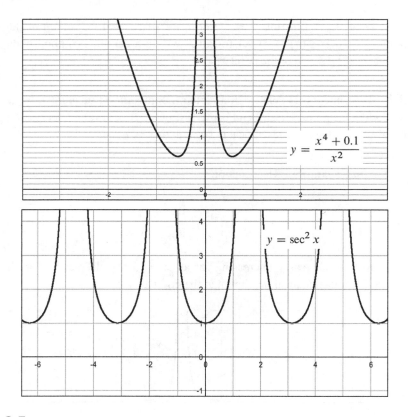

3.5 For a continuous function there is always a strict local maximum between any two local minima.

Counterexample The continuous function below does not have a strict local maximum between its two local minima.

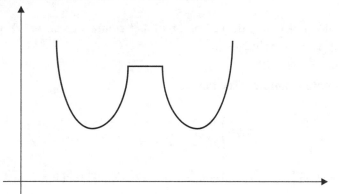

Comment   A function $y = f(x)$ has a strict local maximum at the point $x = a$ if $f(a) > f(x)$ for all x within a certain neighbourhood $(a - \delta, a + \delta)$, $\delta > 0$ of the point $x = a$.

3.6   If a function is defined in a certain neighborhood of point $x = a$ (including the point a itself) and is increasing for all $x < a$ and decreasing for all $x > a$, then there is a local maximum at $x = a$.

Counterexample   The function

$$y = \begin{cases} \frac{1}{(x-3)^2}, & \text{if } x \neq 3 \\ 1, & \text{if } x = 3 \end{cases}$$

is defined for all real x, increasing for all $x < 3$ and decreasing for all $x > 3$, but it has no local maximum at the point $x = 3$.

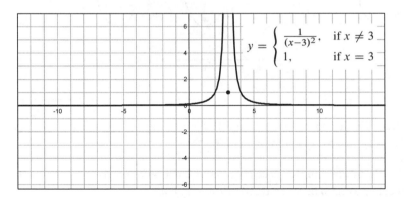

3.7   If a function is defined on $[a, b]$ and continuous on (a, b), then it takes its extreme values on $[a, b]$.

Counterexample   The function

$$y = \begin{cases} \tan x, & \text{if } x \in \left(-\frac{\pi}{2}, \frac{\pi}{2}\right) \\ 0, & \text{if } x = \pm\frac{\pi}{2} \end{cases}$$

is defined on $\left[-\frac{\pi}{2}, \frac{\pi}{2}\right]$ and continuous on $\left(-\frac{\pi}{2}, \frac{\pi}{2}\right)$, but it has no extreme values on $\left[-\frac{\pi}{2}, \frac{\pi}{2}\right]$.

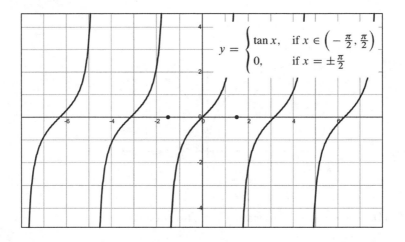

3.8 Every continuous and bounded function on $(-\infty, \infty)$ takes on its extreme values.

Counterexample The function $f(x) = \tan^{-1}(x)$ is continuous and bounded on $(-\infty, \infty)$, but takes no extreme values.

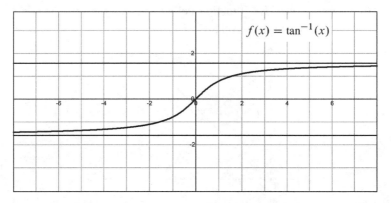

3.9 If a function $f(x)$ is continuous on $[a, b]$, the tangent line exists at all points on its graph and $f(a) = f(b)$, then there is a point c in (a, b) such that the tangent line at the point $(c, f(c))$ is horizontal.

Counterexample The function $f(x)$ below is continuous on $[a, b]$, the tangent line exists at all points on the graph, and $f(a) = f(b)$, but there is no point c in (a, b) such that the tangent line at the point $(c, f(c))$ is horizontal.

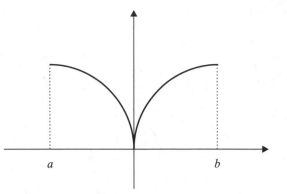

3.10   If on the closed interval $[a, b]$ a function is:

a) bounded;

b) takes its maximum and minimum values;

c) takes all its values between the maximum and minimum values;

then this function is continuous on $[a, b]$.

Counterexample   The function below satisfies the three conditions above, but it is not continuous on $[a, b]$.

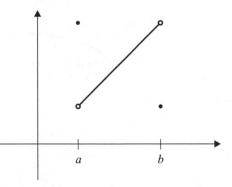

3.11   If on the closed interval $[a, b]$ a function is:

a) bounded;

b) takes its maximum and minimum values;

c) takes all its values between the maximum and minimum values;

then this function is continuous at some points or subintervals on $[a, b]$.

Counterexample   The function below satisfies all three conditions above, but it is discontinuous at *every* point on $[-1, 1]$. It is impossible to draw the graph of the function $f(x)$, but the sketch below gives an idea of its behavior.

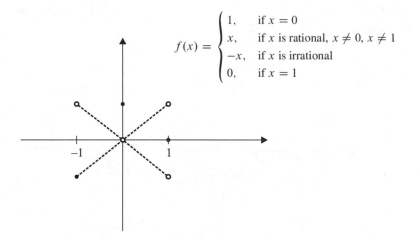

$$f(x) = \begin{cases} 1, & \text{if } x = 0 \\ x, & \text{if } x \text{ is rational, } x \neq 0, x \neq 1 \\ -x, & \text{if } x \text{ is irrational} \\ 0, & \text{if } x = 1 \end{cases}$$

3.12   If a function is continuous on $[a, b]$, then it cannot take its absolute maximum or minimum value infinitely many times.

Counterexample   The function below takes its absolute maximum value ($= 3$) and its absolute minimum value ($= 1$) an infinite number of times on the interval $[1, 4]$.

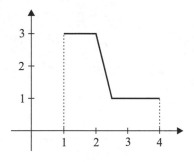

3.13   If a function $f(x)$ is defined on $[a, b]$ and $f(a) \times f(b) < 0$, then there is some point $c \in (a, b)$ such that $f(c) = 0$.

Counterexample   The function

$$f(x) = \begin{cases} \frac{1}{x}, & \text{if } x \neq 0 \\ 1, & \text{if } x = 0 \end{cases}$$

is defined on $[-1, 1]$ and $f(-1) \times f(1) = (-1) \times (1) = -1 < 0$, but there is no point c on $[-1, 1]$ such that $f(c) = 0$.

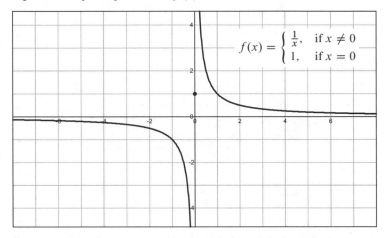

3.14 If a function $f(x)$ is defined on $[a, b]$ and continuous on (a, b), then for any $N \in \big(f(a), f(b)\big)$ there is some point $c \in (a, b)$ such that $f(c) = N$.

Counterexample   The function below is defined on $[a, b]$ and continuous on (a, b), but for any $N \in \big(f(a), f(b)\big)$ there is no point $c \in (a, b)$ such that $f(c) = N$.

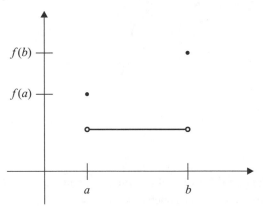

3.15 If a function is discontinuous at every point in its domain, then the square and the absolute value of this function cannot be continuous.

Counterexample The function

$$f(x) = \begin{cases} 1, & \text{if } x \text{ is rational} \\ -1, & \text{if } x \text{ is irrational} \end{cases}$$

is discontinuous at every point in its domain, but both the square and the absolute value

$$f^2(x) = \left| f(x) \right| = 1$$

are continuous. It is impossible to draw the graph of the function $y = f(x)$, but the sketch below gives an idea of its behavior.

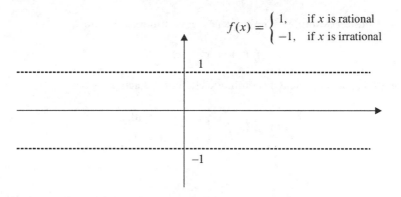

$$f(x) = \begin{cases} 1, & \text{if } x \text{ is rational} \\ -1, & \text{if } x \text{ is irrational} \end{cases}$$

3.16 A function cannot be continuous at only one point in its domain and discontinuous everywhere else.

Counterexample The function

$$g(x) = \begin{cases} x, & \text{if } x \text{ is rational} \\ -x, & \text{if } x \text{ is irrational} \end{cases}$$

is continuous at the point $x = 0$ and discontinuous at all other points on $\mathbb{R}$. It is impossible to draw the graph of the function $y = g(x)$, but the sketch below gives an idea of its behavior.

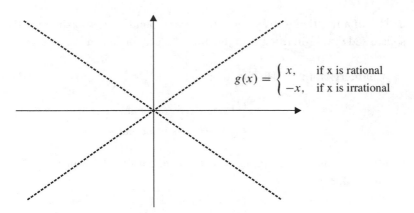

$$g(x) = \begin{cases} x, & \text{if x is rational} \\ -x, & \text{if x is irrational} \end{cases}$$

3.17 A sequence of continuous functions on $[a, b]$ always converges to a continuous function on $[a, b]$.

Counterexample The sequence of continuous functions

$$f_n(x) = x^n, \quad n \in N$$

on $[0, 1]$ converges to a discontinuous function when $n \to \infty$:

$$\lim_{n \to \infty} f_n(x) = \begin{cases} 0, & \text{if } x \in [0, 1) \\ 1, & \text{if } x = 1. \end{cases}$$

4

Differential Calculus

4.1 If both functions $f(x)$ and $g(x)$ are differentiable and $f(x) > g(x)$ on the interval (a, b), then $f'(x) > g'(x)$ on (a, b).

Counterexample Both functions $f(x)$ and $g(x)$ below are differentiable and $f(x) > g(x)$ on the interval (a, b), but $f'(x) < g'(x)$ on (a, b).

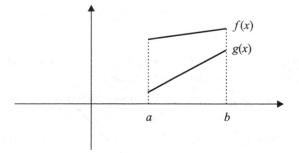

4.2 If a nonlinear function is differentiable and monotone on $(0, \infty)$, then its derivative is also monotone on $(0, \infty)$.

Counterexample The nonlinear function $y = x + \sin x$ is differentiable and monotone on $(0, \infty)$, but its derivative $y' = 1 + \cos x$ is not monotone on $(0, \infty)$.

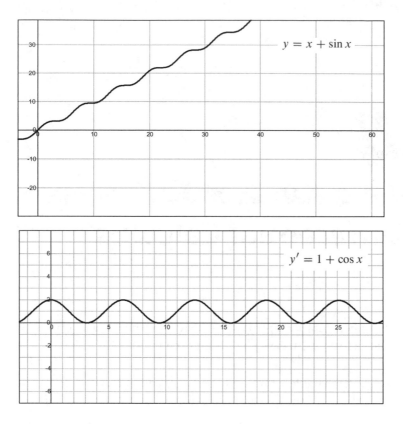

4.3 If a function is continuous at a point, then it is differentiable at that point.

Counterexample The function $y = |x|$ is continuous at the point $x = 0$, but it is not differentiable at that point.

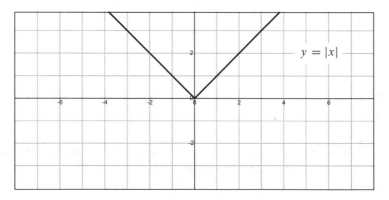

4.4   If a function is continuous on $\mathbb{R}$ and the tangent line exists at every point on its graph, then the function is differentiable at every point on $\mathbb{R}$.

   Counterexample   The function $y = \sqrt[3]{x^2}$ is continuous on $\mathbb{R}$ and the tangent line exists at any point on its graph, but the function is not differentiable at the point $x = 0$.

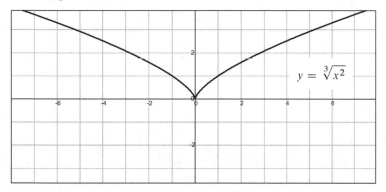

4.5   If a function is continuous on the interval (a, b) and its graph is a *smooth* curve (no sharp corners) on that interval, then the function is differentiable at every point on (a, b).

   Counterexample

a) The function $y = \sqrt[3]{x}$ is continuous on $\mathbb{R}$ and its graph is a smooth curve (no sharp corners), but it is not differentiable at the point $x = 0$.

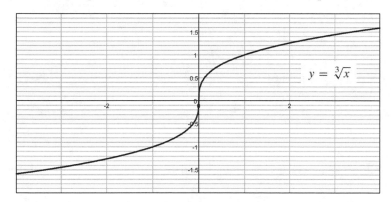

b) The function below is continuous on $\mathbb{R}$ and its graph is a smooth curve (no sharp corners), but it is nondifferentiable at infinitely many points on $\mathbb{R}$.

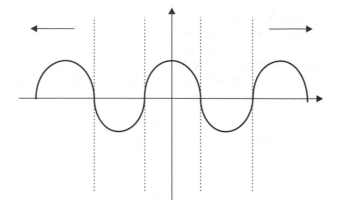

4.6   If the derivative of a function is zero at a point, then the function is neither increasing nor decreasing at this point.

Counterexample   The derivative of the function $y = x^3$ is zero at the point $x = 0$, but the function is increasing at this point.

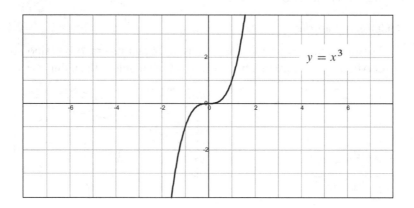

4.7   If a function is differentiable and decreasing on (a, b), then its derivative is negative on (a, b).

Counterexample   The function $y = -x^3$ is differentiable and decreasing on $\mathbb{R}$, but its derivative is zero at the point $x = 0$.

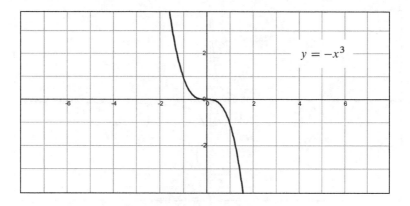

4.8 If a function is continuous and decreasing on (a, b), then its derivative is nonpositive on (a, b).

Counterexample The function below is continuous and decreasing on $\mathbb{R}$, but its derivative does not exist at the point $x = a$.

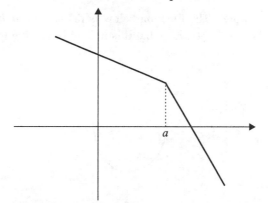

4.9 If a function has a positive derivative at every point in its domain, then the function is increasing everywhere in its domain.

Counterexample The derivative of the function $y = -\frac{1}{x}$ $(x \neq 0)$ is

$$y' = \frac{1}{x^2},$$

which is positive for all $x \neq 0$.

According to the definition, a function is increasing in its domain if for any x_1, x_2 from its domain with $x_1 < x_2$ it follows that $f(x_1) < f(x_2)$. If we take $x_1 = -1$ and $x_2 = 1$ $(x_1 < x_2)$ it follows that $f(x_1) > f(x_2)$.

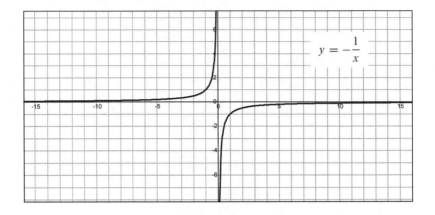

4.10   If a function $f(x)$ is defined on $[a, b]$ and has a local maximum at the point $c \in (a, b)$, then in a sufficiently small neighborhood of the point $x = c$, the function is increasing for all $x < c$ and decreasing for all $x > c$.

Counterexample   The function below is defined on $[a, b]$ and has a maximum at the point $c \in (a, b)$, but it is neither increasing for all $x < c$ nor decreasing for all $x > c$.

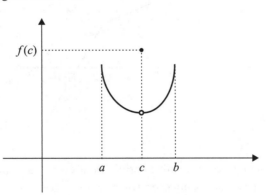

Comment   The definition of a local maximum requires neither differentiability nor continuity of a function at the point of interest: A function $y = f(x)$ has a local maximum at the point $x = c$ if $f(c) > f(x)$ for all x within a certain neighborhood $(c - \delta, c + \delta)$, $\delta > 0$ of the point $x = c$.

4.11   If a function $f(x)$ is differentiable for all real x and $f(0) = f'(0) = 0$, then $f(x) = 0$ for all real x.

Counterexample Both the function $y = x^2$ and its derivative $y' = 2x$ equal zero at the point $x = 0$, but the function is not zero for all real x.

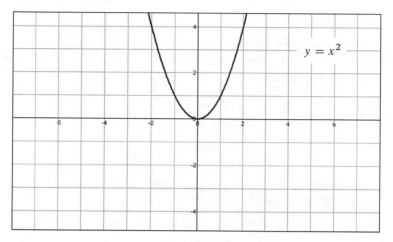

4.12 If a function $f(x)$ is differentiable on the interval (a, b) and takes both positive and negative values on (a, b), then its absolute value $|f(x)|$ is not differentiable at the point(s) where $f(x) = 0$, e.g., $|f(x)| = |x|$ or $|f(x)| = |\sin x|$.

Counterexample The function $y = x^3$ is differentiable on $\mathbb{R}$ and takes both positive and negative values, but its absolute value $|y| = |x^3|$ is differentiable at the point $x = 0$ where the function equals zero.

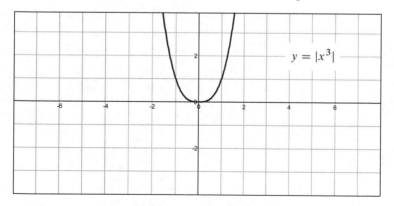

Comment The statement is true if its conclusion is: "...then its absolute value $|f(x)|$ is not differentiable at the points where $f(x) = 0$ and $f'(x) \neq 0$."

4.13   If both functions $f(x)$ and $g(x)$ are differentiable on the interval (a, b) and intersect somewhere on (a, b), then the function max $\{f(x), g(x)\}$ is not differentiable at the point(s) where $f(x) = g(x)$.

Counterexample   The function max$\{x^3, x^4\}$ on $(-1, 1)$ is differentiable at the point $x = 0$ where the functions $y = x^3$ and $y = x^4$ intersect.

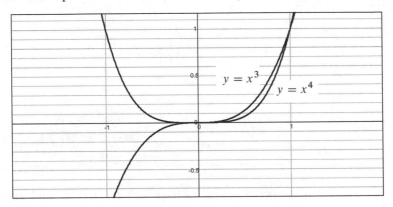

Comment   The statement is true if its conclusion is: "...then the function max $\{f(x), g(x)\}$ is not differentiable at the point(s) where $f(x) = g(x)$ and $f'(x) \neq g'(x)$."

4.14   If a function is twice-differentiable at a local maximum (minimum) point, then its second derivative is negative (positive) at that point.

Counterexample   The function $y = -x^4$ is twice-differentiable at its maximum point $x = 0$, but the second derivative is zero at this point.

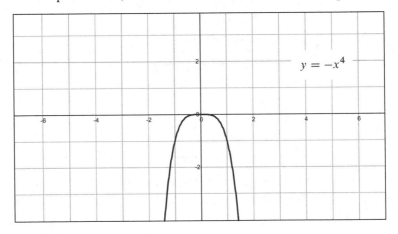

The function $y = x^4$ is twice-differentiable at its minimum point $x = 0$, but the second derivative is zero at that point:

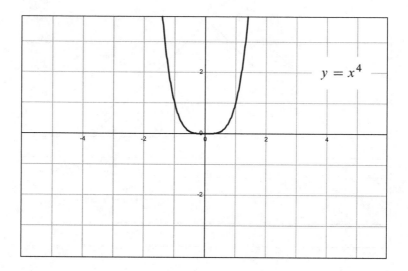

4.15 If both functions $f(x)$ and $g(x)$ are not differentiable at $x = a$, then $f(x) + g(x)$ is also not differentiable at $x = a$.

Counterexample Both functions $f(x) = |x|$ and $g(x) = -|x| + 1$ are not differentiable at $x = 0$, but $f(x) + g(x) = 1$ is differentiable at any x including $x = 0$.

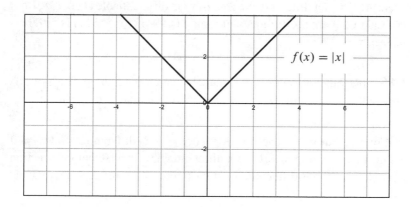

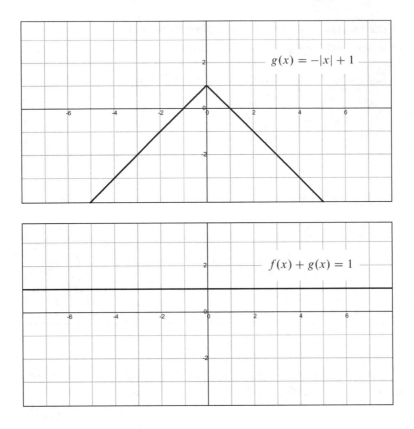

Comment   More generally, let

$$f(x) = A(x) \quad \text{and} \quad g(x) = B(x) - A(x),$$

where $A(x)$ is not differentiable and $B(x)$ is differentiable at $x = a$. Both $f(x)$ and $g(x)$ are not differentiable, but $f(x) + g(x) = B(x)$ is differentiable at $x = a$.

4.16   If a function $f(x)$ is differentiable and a function $g(x)$ is not differentiable at $x = a$, then $f(x)\,g(x)$ is not differentiable at $x = a$.

Counterexample   The function $f(x) = x$ is differentiable at $x = 0$ and the function $g(x) = |x|$ is not differentiable at $x = 0$, but the function $f(x)\,g(x) = x|x|$ is differentiable at the point $x = 0$ (the derivative equals zero).

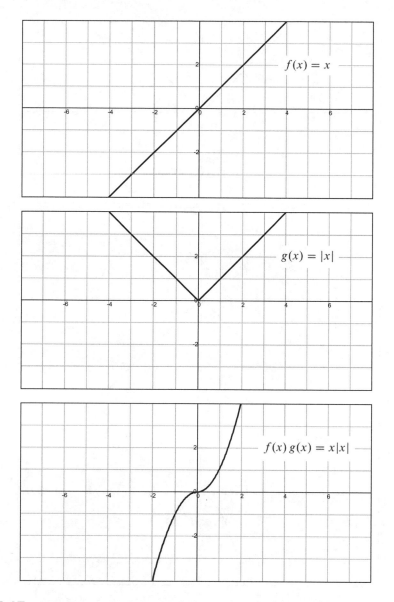

4.17 If both functions $f(x)$ and $g(x)$ are not differentiable at $x = a$, then $f(x) g(x)$ is also not differentiable at $x = a$.

Counterexample Both functions $f(x) = |x|$ and $g(x) = -|x|$ are not differentiable at the point $x = 0$, but the function $f(x) g(x) = -|x|^2 = -x^2$ is differentiable at $x = 0$.

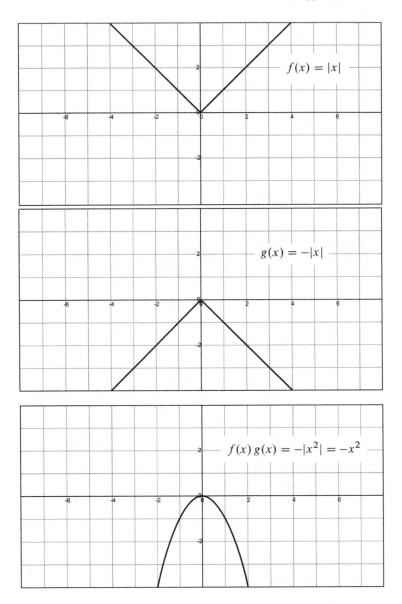

4.18 If a function $g(x)$ is differentiable at $x = a$ and a function $f(x)$ is not differentiable at $g(a)$, then the function $F(x) = f(g(x))$ is not differentiable at $x = a$.

Counterexample   The function $g(x) = x^2$ is differentiable at $x = 0$, and the function $f(x) = |x|$ is not differentiable at $g(0) = 0$, but the function

$$F(x) = f(g(x)) = |x^2| = x^2$$

is differentiable at $x = 0$.

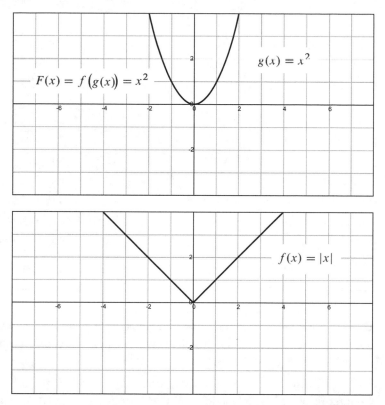

4.19   If a function $g(x)$ is not differentiable at $x = a$ and a function $f(x)$ is differentiable at g(a), then the function $F(x) = f(g(x))$ is not differentiable at $x = a$.

Counterexample   The function $g(x) = |x|$ is not differentiable at $x = 0$, the function $f(x) = x^2$ is differentiable at $g(0) = 0$, but the function

$$F(x) = f(g(x)) = |x|^2 = x^2$$

is differentiable at $x = 0$.

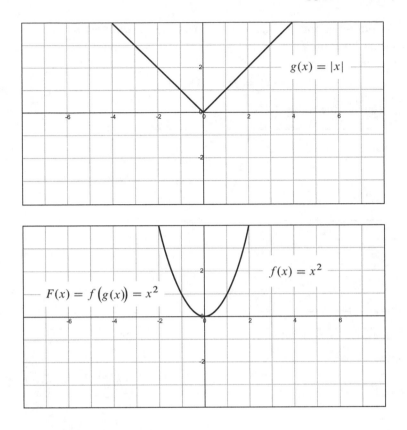

4.20 If a function $g(x)$ is not differentiable at $x = a$ and a function $f(x)$ is not differentiable at $g(a)$, then the function $F(x) = f\big(g(x)\big)$ is not differentiable at $x = a$.

Counterexample   The function

$$g(x) = \frac{2}{3}x - \frac{1}{3}|x|$$

is not differentiable at $x = 0$ and the function $f(x) = 2x + |x|$ is not differentiable at $g(0) = 0$, but the function

$$F(x) = f\big(g(x)\big) = 2\left(\frac{2}{3}x - \frac{1}{3}|x|\right) + \left|\frac{2}{3}x - \frac{1}{3}|x|\right| = x$$

is differentiable at $x = 0$.

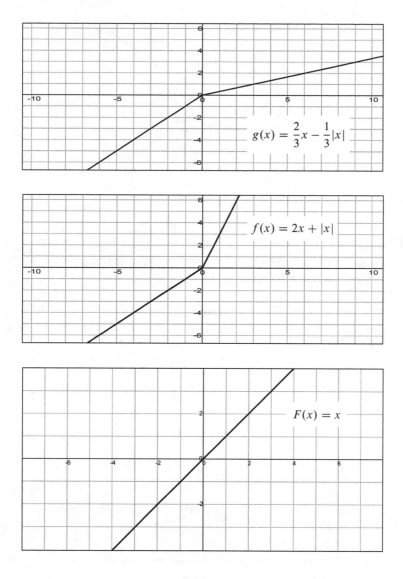

$$g(x) = \frac{2}{3}x - \frac{1}{3}|x|$$

$$f(x) = 2x + |x|$$

$$F(x) = x$$

4.21 If a function $f(x)$ is defined on $[a, b]$, differentiable on (a, b) and $f(a) = f(b)$, then there exists a point $c \in (a, b)$ such that $f'(c) = 0$.

Counterexample The function below is defined on $[a, b]$, differentiable on (a, b) and $f(a) = f(b)$, but there is no such point $c \in (a, b)$ that $f'(c) = 0$.

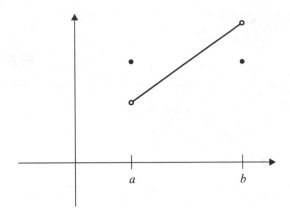

4.22   If a function is twice-differentiable in a certain neighborhood around $x = a$ and its second derivative is zero at that point, then the point $(a, f(a))$ is a point of inflection for the graph of the function.

Counterexample   The function $y = x^4$ is twice-differentiable on $\mathbb{R}$ and its second derivative is zero at the point $x = 0$, but the point $(0, 0)$ is not a point of inflection.

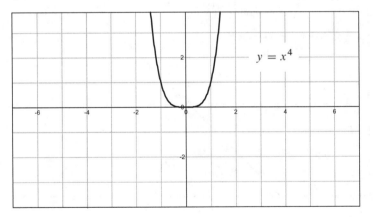

4.23   If a function $f(x)$ is differentiable at the point $x = a$ and the point $(a, f(a))$ is a point of inflection on the function's graph, then the second derivative is zero at that point.

Counterexample   The function $y = x|x|$ is differentiable at $x = 0$ and the point $(0, 0)$ is a point of inflection, but the second derivative does not exist at $x = 0$.

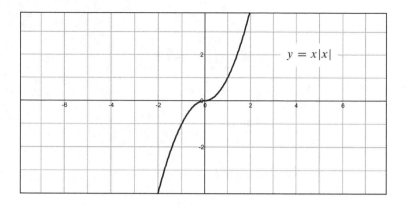

4.24 If both functions $f(x)$ and $g(x)$ are differentiable on $\mathbb{R}$, then to evaluate the limit

$$\lim_{x \to \infty} \frac{f(x)}{g(x)}$$

in the indeterminate form of type $\left[\frac{\infty}{\infty}\right]$ we can use the following rule:

$$\lim_{x \to \infty} \frac{f(x)}{g(x)} = \lim_{x \to \infty} \frac{f'(x)}{g'(x)}.$$

Counterexample   If we use the above "rule" to find the limit

$$\lim_{x \to \infty} \frac{6x + \sin x}{2x + \sin x},$$

then

$$\lim_{x \to \infty} \frac{6x + \sin x}{2x + \sin x} = \left[\frac{\infty}{\infty}\right] = \lim_{x \to \infty} \frac{6 + \cos x}{2 + \cos x}$$

is undefined (see the graph below).

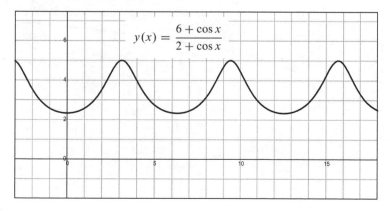

But the limit

$$\lim_{x \to \infty} \frac{6x + \sin x}{2x + \sin x}$$

exists and equals 3:

$$\lim_{x \to \infty} \frac{6x + \sin x}{2x + \sin x} = \lim_{x \to \infty} \frac{6 + \frac{\sin x}{x}}{2 + \frac{\sin x}{x}} = 3.$$

Comment   To make the statement correct we need to add "if the limit

$$\lim_{x \to \infty} \frac{f'(x)}{g'(x)}$$

exists or equals $\pm\infty$." This is the well-known l'Hospital's Rule for limits.

4.25   If a function $f(x)$ is differentiable on (a, b) and $\lim_{x \to a+} f'(x) = \infty$, then $\lim_{x \to a+} f(x) = \infty$.

Counterexample   The function $y = \sqrt[3]{x}$ is differentiable on $(0, 1)$ and

$$\lim_{x \to 0+} y'(x) = \lim_{x \to 0+} \frac{1}{3\sqrt[3]{x^2}} = \infty,$$

but

$$\lim_{x \to 0+} y(x) = \lim_{x \to 0+} \sqrt[3]{x} = 0.$$

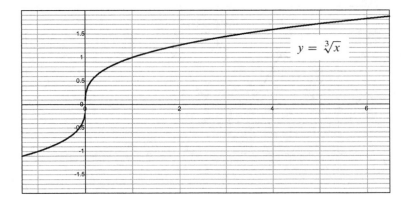

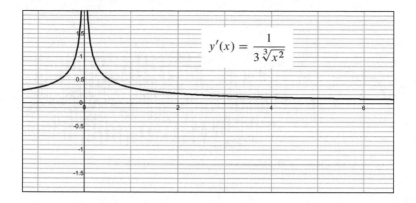

$$y'(x) = \frac{1}{3\sqrt[3]{x^2}}$$

4.26 If a function $f(x)$ is differentiable on $(0, \infty)$ and $\lim_{x \to \infty} f(x)$ exists, then $\lim_{x \to \infty} f'(x)$ also exists.

Counterexample   The function

$$f(x) = \frac{\sin(x^2)}{x}$$

is differentiable on $(0, \infty)$ and

$$\lim_{x \to \infty} \frac{\sin(x^2)}{x} = 0,$$

but

$$\lim_{x \to \infty} f'(x) = \lim_{x \to \infty} \frac{2x^2 \cos(x^2) - \sin(x^2)}{x^2}$$

does not exist.

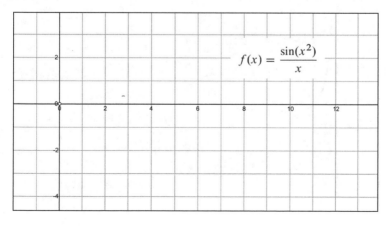

$$f(x) = \frac{\sin(x^2)}{x}$$

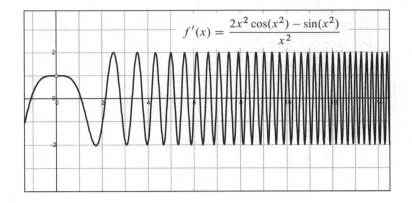

4.27 If a function $f(x)$ is differentiable and bounded on $(0, \infty)$ and $\lim_{x \to \infty} f'(x)$ exists, then $\lim_{x \to \infty} f(x)$ also exists.

Counterexample The function $f(x) = \cos(\ln x)$ is differentiable and bounded on $(0, \infty)$ and the limit of its derivative exists:

$$\lim_{x \to \infty} f'(x) = \lim_{x \to \infty} -\frac{\sin(\ln x)}{x} = 0.$$

However, the limit of the function $\lim_{x \to \infty} \cos(\ln x)$ does not exist.

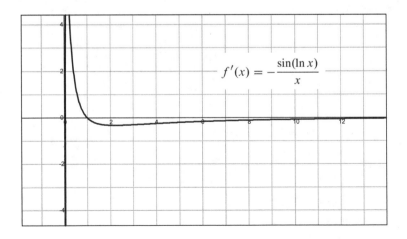

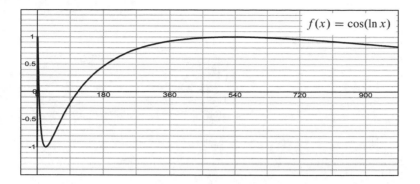

Comment   Obviously, the function $\cos(x)$ oscillates between -1 and 1 as $x \to \infty$ and the limit $\lim_{x \to \infty} \cos x$ does not exist. The log function $\ln x$ tends to infinity as $x \to \infty$ so the function $f(x) = \cos(\ln x)$ also oscillates between -1 and 1 and the limit $\lim_{x \to \infty} \cos(\ln x)$ does not exist.

4.28   If a function $f(x)$ is differentiable at the point $x = a$, then its derivative is continuous at $x = a$.

Counterexample   The function

$$f(x) = \begin{cases} x^2 \sin \frac{1}{x}, & \text{if } x \neq 0 \\ 0, & \text{if } x = 0 \end{cases}$$

is differentiable at $x = 0$, but its derivative

$$f'(x) = \begin{cases} 2x \sin \frac{1}{x} - \cos \frac{1}{x}, & \text{if } x \neq 0 \\ 0, & \text{if } x = 0 \end{cases}$$

is discontinuous at $x = 0$.

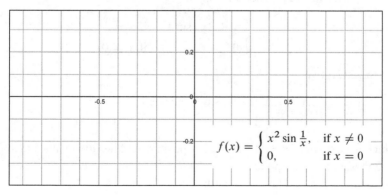

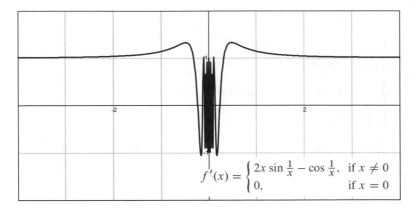

$$f'(x) = \begin{cases} 2x \sin \frac{1}{x} - \cos \frac{1}{x}, & \text{if } x \neq 0 \\ 0, & \text{if } x = 0 \end{cases}$$

Comment To show that the derivative of $f(x)$ is 0 at the point 0 we use the definition of the derivative:

$$f'(0) = \lim_{h \to 0} \frac{f(0+h) - f(0)}{h} = \lim_{h \to 0} \frac{h^2 \sin(1/h) - 0}{h}$$

$$= \lim_{h \to 0} \left(h \sin \frac{1}{h} \right).$$

Since $-1 \leq \sin \frac{1}{h} \leq 1$ then $-h \leq h \sin \frac{1}{h} \leq h$ when $h > 0$ and $h \leq h \sin \frac{1}{h} \leq -h$ when $h < 0$. Applying the Squeeze Theorem we get that

$$\lim_{h \to 0} \left(h \sin \frac{1}{h} \right) = 0.$$

4.29 If the derivative of a function $f(x)$ is positive at the point $x = a$, then there exists a neighborhood about $x = a$ where the function is increasing.

Counterexample The function

$$f(x) = \begin{cases} x + 2x^2 \sin \frac{1}{x}, & \text{if } x \neq 0 \\ 0, & \text{if } x = 0 \end{cases}$$

has the derivative

$$f'(x) = \begin{cases} 1 + 4x \sin \frac{1}{x} - 2 \cos \frac{1}{x}, & \text{if } x \neq 0 \\ 1, & \text{if } x = 0 \end{cases}$$

which is positive at $x = 0$, but it takes positive and negative values in any neighborhood of the point $x = 0$. This means the function $f(x)$ is not monotone in any neighborhood of the point $x = 0$.

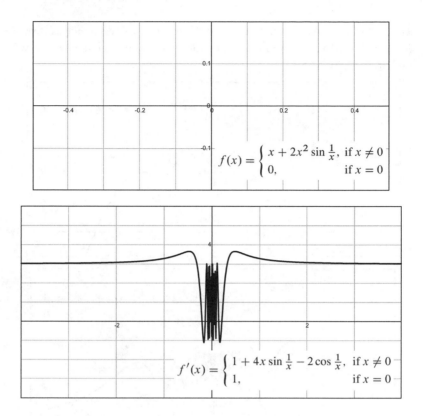

$$f(x) = \begin{cases} x + 2x^2 \sin \frac{1}{x}, & \text{if } x \neq 0 \\ 0, & \text{if } x = 0 \end{cases}$$

$$f'(x) = \begin{cases} 1 + 4x \sin \frac{1}{x} - 2 \cos \frac{1}{x}, & \text{if } x \neq 0 \\ 1, & \text{if } x = 0 \end{cases}$$

4.30 If a function $f(x)$ is continuous on (a, b) and has a local maximum at the point $c \in (a, b)$, then in a sufficiently small neighborhood of the point $x = c$ the function is increasing for all $x < c$ and decreasing for all $x > c$.

Counterexample   The function

$$f(x) = \begin{cases} 2 - x^2(2 + \sin \frac{1}{x}), & \text{if } x \neq 0 \\ 2, & \text{if } x = 0 \end{cases}$$

is continuous on $\mathbb{R}$. Since $x^2(2 + \sin \frac{1}{x})$ is positive for all $x \neq 0$, then $2 > 2 - x^2(2 + \sin \frac{1}{x})$. Therefore the function $y = f(x)$ has a local maximum at the point $x = 0$. But it is neither increasing for all $x < 0$ nor decreasing for all $x > 0$ in any neighborhood of the point $x = 0$. To show this we can find the derivative

$$f'(x) = -4x - 2x \sin \frac{1}{x} + \cos \frac{1}{x}; \quad x \neq 0.$$

The derivative takes both positive and negative values in any interval $(-\delta, 0) \cup (0, \delta)$ and therefore the function is not monotone in any interval $(-\delta, 0) \cup (0, \delta)$, where $\delta > 0$.

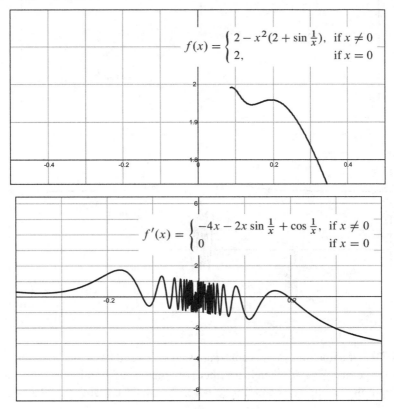

$$f(x) = \begin{cases} 2 - x^2(2 + \sin \frac{1}{x}), & \text{if } x \neq 0 \\ 2, & \text{if } x = 0 \end{cases}$$

$$f'(x) = \begin{cases} -4x - 2x \sin \frac{1}{x} + \cos \frac{1}{x}, & \text{if } x \neq 0 \\ 0 & \text{if } x = 0 \end{cases}$$

4.31 If a function $f(x)$ is differentiable at the point $x = a$, then there is a certain neighborhood of the point $x = a$ where the derivative of the function $f(x)$ is bounded.

Counterexample The function

$$f(x) = \begin{cases} x^2 \sin \frac{1}{x^2}, & \text{if } x \neq 0 \\ 0, & \text{if } x = 0 \end{cases}$$

is differentiable at the point $x = 0$. Its derivative is

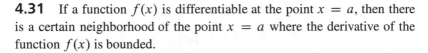

$$f'(x) = \begin{cases} 2x \sin \frac{1}{x^2} - \frac{2}{x} \cos \frac{1}{x^2}, & \text{if } x \neq 0 \\ 0, & \text{if } x = 0 \end{cases}.$$

The derivative of the function $y = f(x)$ is unbounded in any neighborhood of the point $x = 0$.

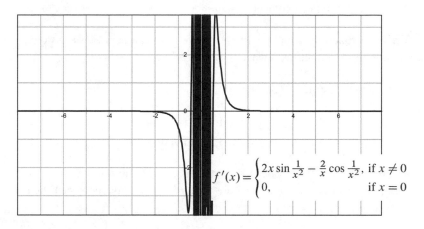

$$f'(x) = \begin{cases} 2x \sin \frac{1}{x^2} - \frac{2}{x} \cos \frac{1}{x^2}, & \text{if } x \neq 0 \\ 0, & \text{if } x = 0 \end{cases}$$

4.32   If a function $f(x)$ in every neighborhood of the point $x = a$ has points where $f'(x)$ does not exist, then $f'(a)$ does not exist.

Counterexample   The function

$$f(x) = \begin{cases} x^2 \left| \cos \frac{\pi}{x} \right|, & \text{if } x \neq 0 \\ 0, & \text{if } x = 0 \end{cases}$$

in any neighborhood of the point $x = 0$ has points where $f'(x)$ does not exist, however $f'(0) = 0$.

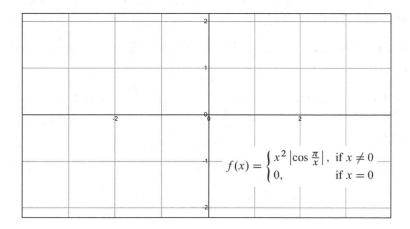

$$f(x) = \begin{cases} x^2 \left| \cos \frac{\pi}{x} \right|, & \text{if } x \neq 0 \\ 0, & \text{if } x = 0 \end{cases}$$

Comment   To show that the derivative of $f(x)$ is 0 at the point 0 we can use the definition of the derivative and the Squeeze Theorem in a similar way as we did in the comment to statement 4.28 from this section.

4.33   A function cannot be differentiable only at one point in its domain and nondifferentiable everywhere else in its domain.

Counterexample   The function

$$y = \begin{cases} x^2, & \text{if } x \text{ is rational} \\ 0, & \text{if } x \text{ is irrational} \end{cases}$$

is defined for all real x and differentiable only at the point $x = 0$. It is impossible to draw the graph of the function $y(x)$, but the sketch below gives an idea of its behavior.

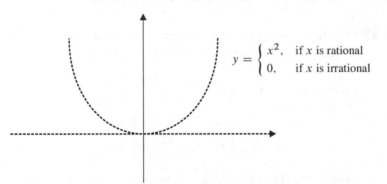

$$y = \begin{cases} x^2, & \text{if } x \text{ is rational} \\ 0, & \text{if } x \text{ is irrational} \end{cases}$$

Comment   To find the derivative of $y(x)$ at the point 0 one can use the definition of the derivative.

4.34   If a function is continuous on (a, b), then it is differentiable at some points on (a, b).

Counterexample   The Weierstrass function named after the great German mathematician Karl Weierstrass (1815–1897) can be defined as:

$$f(x) = \sum_{n=0}^{\infty} \left(\frac{1}{2}\right)^n \cos(3^n x).$$

It is continuous, but nondifferentiable everywhere on $\mathbb{R}$. If we take the first seven terms in the sum, we can begin to visualize the function:

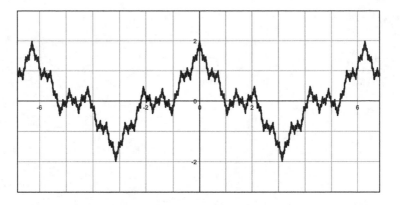

Comments The Weierstrass function is the first known fractal. An exercise to explore the Weierstrass function is given in (Smith & Minton, 2002) on p. 176. More about the Weierstrass function can be found in (Brabenec, 2004) and (Bressoud, 1995). Another example of a continuous curve that has a sharp corner at every point is the Koch snowflake named after the Swedish mathematician Helge von Koch (1870–1924). We start with an equilateral triangle and build the line segments on each side according to a simple rule and repeat this process infinitely many times. The resulting curve is called the Koch curve and it forms the so-called Koch snowflake. The first four iterations are shown below:

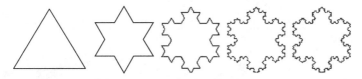

5

Integral Calculus

5.1   If the function $F(x)$ is an antiderivative of a function $f(x)$, then

$$\int_a^b f(x)dx = F(b) - F(a).$$

Counterexample   The function

$$F(x) = \ln|x|$$

is an antiderivative of the function

$$f(x) = \frac{1}{x},$$

but the (improper) integral

$$\int_{-1}^1 \frac{1}{x}\, dx$$

does not exist.

Comments   To make the statement true we need to add that the function $f(x)$ must be continuous on $[a, b]$.

5.2   If a function $f(x)$ is continuous on $[a, b]$, then the area enclosed by the graph of $y = f(x)$, $y = 0$, $x = a$ and $x = b$ numerically equals

$$\int_a^b f(x)\, dx.$$

Counterexample   For any continuous function $f(x)$ that takes only negative values on $[a, b]$ the integral

$$\int_a^b f(x)\,dx$$

is negative, therefore the area enclosed by the graph of $f(x)$, $y = 0$, $x = a$ and $x = b$ is numerically equal to

$$-\int_a^b f(x)\,dx,$$

or

$$\left| \int_a^b f(x)\,dx \right|.$$

5.3   If

$$\int_a^b f(x)\,dx \geq 0,$$

then $f(x) \geq 0$ for all $x \in [a, b]$.

Counterexample

$$\int_{-1}^2 x\,dx = \frac{3}{2} > 0,$$

but the function $y = x$ takes both positive and negative values on $[-1, 2]$.

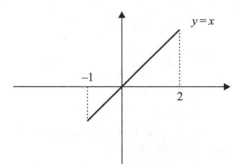

5.4   If $f(x)$ is a continuous function and k is any constant, then:

$$\int k f(x)\,dx = k \int f(x)\,dx.$$

Counterexample   If $k = 0$, then the left-hand side is:

$$\int 0 f(x)\, dx = \int 0\, dx = C,$$

where C is an arbitrary constant. The right-hand side is:

$$0 \int f(x)\, dx = 0.$$

This indicates that C is always equal to zero, but this contradicts the nature of an arbitrary constant.

Comment   The property is valid only for nonzero values of the constant k.

5.5   A plane figure of infinite area rotated around an axis always produces a solid of revolution of infinite volume.

Counterexample   The figure enclosed by the graph of the function

$$y = \frac{1}{x},$$

the x-axis and the straight line $x = 1$ is rotated about the x-axis.

The area is infinite:

$$\int_1^{\infty} \frac{1}{x}\, dx = \lim_{b \to \infty} (\ln b - \ln 1) = \infty$$

(square units), but the volume is finite:

$$\pi \int_1^{\infty} \frac{1}{x^2}\, dx = -\pi \lim_{b \to \infty} \left(\frac{1}{b} - 1 \right) = \pi$$

(cubic units).

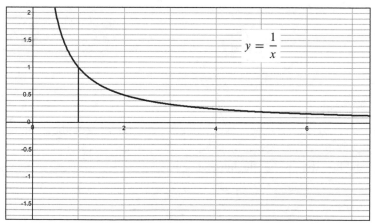

Z axis

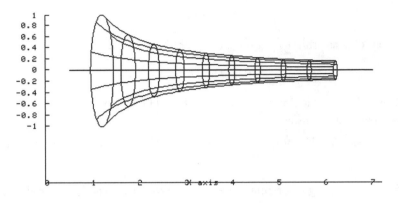

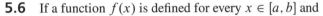

5.6   If a function $f(x)$ is defined for every $x \in [a, b]$ and

$$\int_a^b |f(x)|\, dx$$

exists, then

$$\int_a^b f(x)\, dx$$

exists.

Counterexample   Again we use a Dirichlet function

$$f(x) = \begin{cases} 1, & \text{if } x \text{ is rational} \\ -1, & \text{if } x \text{ is irrational} \end{cases}.$$

$|f(x)| = 1$ and therefore

$$\int_a^b |f(x)|\, dx = b - a,$$

but

$$\int_a^b f(x)\, dx$$

does not exist. Let us show this using the definition of the definite integral.

Let $[a, b]$ be any closed interval. We divide the interval into n subintervals and find the limit of the integral sums:

$$S = \lim_{\max \Delta x_i \to 0} \sum_{i=0}^{n-1} f(c_i)\Delta x_i.$$

If on each subinterval we choose c_i equal to a rational number, then $S = b - a$. If instead on each subinterval we choose c_i equal to an irrational number, then $S = a - b$. So, the limit of the integral sums depends on the way we choose c_i and for this reason the definite integral of $f(x)$ on $[a, b]$ does not exist.

5.7 If neither of the integrals

$$\int_a^b f(x)\, dx \quad \text{and} \quad \int_a^b g(x)\, dx$$

exist, then the integral

$$\int_a^b (f(x) + g(x))dx$$

does not exist.

Counterexample For the functions

$$f(x) = \begin{cases} 1, & \text{if } x \text{ is rational} \\ -1, & \text{if } x \text{ is irrational} \end{cases} \quad \text{and} \quad g(x) = \begin{cases} -1, & \text{if } x \text{ is rational} \\ 1, & \text{if } x \text{ is irrational} \end{cases}$$

the integrals

$$\int_a^b f(x)\, dx \quad \text{and} \quad \int_a^b g(x)\, dx$$

do not exist (see the previous exercise), but the integral

$$\int_a^b \left(f(x) + g(x)\right) dx$$

exists and equals 0.

5.8 If $\lim_{x \to \infty} f(x) = 0$, then

$$\int_a^\infty f(x)\, dx$$

converges.

Counterexample   The limit

$$\lim_{x \to \infty} \frac{1}{x} = 0,$$

but the integral

$$\int_1^\infty \frac{1}{x}\, dx$$

diverges.

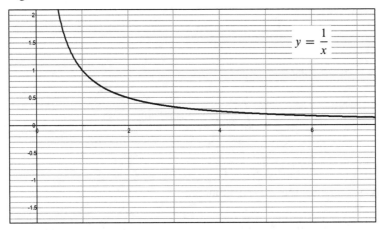

5.9   If the integral

$$\int_a^\infty f(x)\, dx$$

diverges, then the function $f(x)$ is not bounded.

Counterexample   The integral of a nonzero constant

$$\int_a^\infty k\, dx$$

is divergent, but the function $y = k$ is bounded.

5.10   If a function $f(x)$ is continuous and nonnegative for all real x and

$$\sum_{n=1}^\infty f(n)$$

is finite, then

$$\int_1^\infty f(x)\, dx$$

converges.

Counterexample   The function

$$y = |\sin \pi\, x|$$

is continuous and nonnegative for all real x and

$$\sum_{n=1}^{\infty} |\sin \pi\, n| = 0,$$

but

$$\int_{1}^{\infty} |\sin \pi\, x|\ dx$$

diverges.

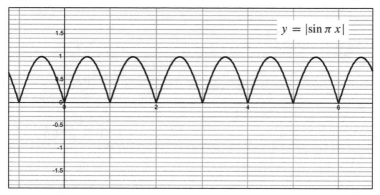

5.11   If both integrals

$$\int_{a}^{\infty} f(x)\, dx \quad \text{and} \quad \int_{a}^{\infty} g(x)\, dx$$

diverge, then the integral

$$\int_{a}^{\infty} \big(f(x) + g(x)\big)\, dx$$

also diverges.

Counterexample   Both integrals

$$\int_{1}^{\infty} \frac{1}{x}\, dx \quad \text{and} \quad \int_{1}^{\infty} \frac{1-x}{x^2}\, dx$$

diverge, but the integral

$$\int_{1}^{\infty} \left(\frac{1}{x} + \frac{1-x}{x^2}\right) dx = \int_{1}^{\infty} \frac{1}{x^2}\, dx$$

converges.

5.12   If a function $f(x)$ is continuous and

$$\int_a^\infty f(x)\,dx$$

converges, then

$$\lim_{x\to\infty} f(x) = 0.$$

Counterexample   It can be shown that the improper integral

$$\int_0^\infty \sin x^2\,dx$$

converges, but

$$\lim_{x\to\infty} \sin x^2$$

does not exist.

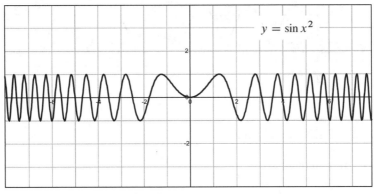

Comment   The function

$$F(x) = \int_0^x \sin t^2\,dt$$

is called the Fresnel function or the Fresnel integral, named after the French physicist Augustin Fresnel (1788–1827). It cannot be evaluated analytically in terms of a finite number of elementary functions. It can be represented only as an (infinite) power series. A good introduction to the Fresnel integral is given in (Stewart, 2001) on pp. 383–384 for illustrating The Fundamental Theorem of Calculus. It can be shown that the improper integral

$$\int_0^\infty \sin x^2\,dx \quad \text{equals} \quad \sqrt{\frac{\pi}{8}}$$

by the methods of complex analysis, but this is beyond the scope of this book. The Fresnel integral has applications in optics and in highway design.

5.13 If a function $f(x)$ is continuous and nonnegative and

$$\int_a^\infty f(x)\,dx$$

converges, then

$$\lim_{x\to\infty} f(x) = 0.$$

Counterexample We use the idea of area. Over every natural $n \geq 2$ we construct triangles of area $1/n^2$ so that the total area equals

$$\sum_{n=a}^\infty \frac{1}{n^2},$$

which is a finite number. The height of each triangle is n and the base is $2/n^3$. The integral

$$\int_a^\infty f(x)\,dx$$

converges since it is numerically equal to the total area

$$\sum_{n=a}^\infty \frac{1}{n^2}.$$

As one can see from the graph below the function (in bold) is continuous and nonnegative, but

$$\lim_{x\to\infty} f(x)$$

does not exist.

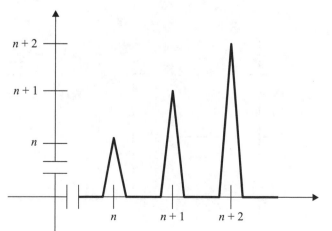

5.14   If a function $f(x)$ is positive and unbounded for all real x, then the integral

$$\int_a^\infty f(x)\,dx$$

diverges.

Counterexample   We use the idea of area. Over every natural n we can construct a rectangle with height n and base $1/n^3$ so the area is $1/n^2$. Then the total area equals

$$\sum_{n=a}^\infty \frac{1}{n^2},$$

which is a finite number. The positive and unbounded function equals n on the interval of length $1/n^3$ around points $x = n$, where n is natural. Since the integral

$$\int_a^\infty f(x)\,dx$$

numerically equals the total area

$$\sum_{n=a}^\infty \frac{1}{n^2}$$

it converges.

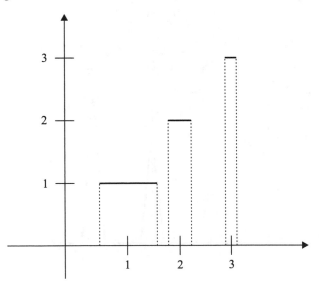

5.15 If a function $f(x)$ is continuous and unbounded for all real x, then the integral

$$\int_0^\infty f(x)\,dx$$

diverges.

Counterexample The function

$$y = x\sin x^4$$

is continuous and unbounded for all real x, but the integral

$$\int_0^\infty x\sin x^4\,dx$$

converges (making the substitution $t = x^2$ yields the Fresnel integral

$$\frac{1}{2}\int_0^\infty \sin t^2\,dt$$

which is convergent—see the comments to statement 5.12).

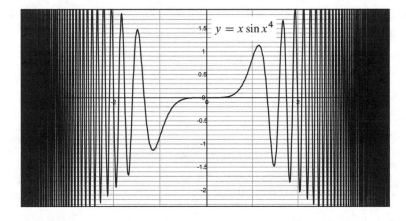

5.16 If a function $f(x)$ is continuous on $[1, \infty)$ and

$$\int_1^\infty f(x)\,dx$$

converges, then

$$\int_1^\infty \left|f(x)\right|\,dx$$

also converges.

Counterexample   The function

$$y = \frac{\sin x}{x}$$

is continuous on $[1, \infty)$ and

$$\int_1^\infty \frac{\sin x}{x}\, dx$$

converges, but

$$\int_1^\infty \left| \frac{\sin x}{x} \right|\, dx$$

diverges.

Comments   The integral of the so-called sinc function $\frac{\sin x}{x}$ is called the sine integral and is denoted by

$$Si(x) = \int_0^x \frac{\sin t}{t}\, dt.$$

It cannot be evaluated analytically in terms of a finite number of elementary functions. One can verify that

$$\int_1^\infty \frac{\sin x}{x}\, dx \le \int_0^\infty \frac{\sin x}{x}\, dx \le \pi \sum_{n=1}^\infty \frac{(-1)^{n+1}}{n}$$

and therefore the improper integral

$$\int_1^\infty \frac{\sin x}{x}\, dx$$

is convergent. It can be shown that it equals $\frac{\pi}{2}$.

5.17   If the integral

$$\int_a^\infty f(x)\, dx$$

converges and a function $g(x)$ is bounded, then the integral

$$\int_a^\infty f(x)g(x)\, dx$$

converges.

Counterexample The integral

$$\int_0^\infty \frac{\sin x}{x}\, dx$$

converges (see the comments to statement 5.16 above) and the function $g(x) = \sin x$ is bounded, but the integral

$$\int_0^\infty \frac{\sin^2 x}{x}\, dx$$

diverges.

Comment Statements 5.10, 5.13 and 5.14 in this chapter were supplied by Alejandro S. Gonzalez-Martin, University La Laguna, Spain.

References

Aaboe, A. (1975) *Episodes From the Early History of Mathematics*. Washington, D.C.: The Mathematical Association of America.

Boyer, C. (1991) *A History of Mathematics*. 2nd edition. Wiley.

Brabenec, R. (2004) *Resources For the Study of Real Analysis*. Washington, D.C.: The Mathematical Association of America.

Bressoud, D. (1995). *A Radical Approach to Real Analysis*. Washington, D.C.: The Mathematical Association of America.

Dunham, W. (2005). *The Calculus Gallery*. Princeton University Press.

Gelbaum, B.R., Olmsted, J.M.H. (1964) *Counterexamples in Analysis*. Holden-Day, Inc., San Francisco.

Gruenwald, N., Klymchuk, S. (2003) Using counter-examples in teaching Calculus. *The New Zealand Mathematics Magazine*. 40(2), 33–41.

Klymchuk, S. (2004) *Counter-Examples in Calculus*. Maths Press, New Zealand.

—— (2005) Counter-examples in teaching/learning of Calculus: Students' performance. *The New Zealand Mathematics Magazine*. 42(1), 31–38.

Richstein, J. (2000) Verifying the Goldbach conjecture up to $4 \cdot 10^{14}$. *Mathematics of Computation*. 70 (236), 1745–1749.

Selden, A., Selden J. (1998) The role of examples in learning mathematics. *MAA Online*: www.maa.org/t_and_l/sampler/rs_5.html.

Smith, R.T., Minton, R.B. (2002) Calculus. 2nd edition. McGraw-Hill.

Stewart, J. (2001) *Calculus: Concepts and Contexts*. 2nd edition. Brooks/Cole, Thomson Learning.

Tall, D. (1991) The psychology of advanced mathematical thinking. In: Tall (ed) *Advanced Mathematical Thinking*. Kluwer: Dordrecht, 3–21.

About the Author

Sergiy Klymchuk is an Associate Professor of the School of Computing and Mathematical Sciences at the Auckland University of Technology, New Zealand. He has 29 years experience teaching university mathematics in different countries. His PhD (1988) was in differential equations and recent research interests are in mathematics education and epidemic modeling. He has more than 140 publications including several books on popular mathematics and science that have been, or are being, published in 11 countries.